Design with Digital Tools

Using New Media Creatively

Mark von Wodtke

McGraw-Hill

New York
San Francisco
Washington, D.C.
Auckland
Bogotá
Caracas
Lisbon
London
Madrid
Mexico City
Milan
Montreal
New Delhi
San Juan
Singapore
Sydney
Tokyo
Toronto

Library of Congress Cataloging-in-Publication Data

Von Wodtke, Mark.
   Design with digital tools : using new media creatively / Mark von Wodtke.
      p.   cm.
   Includes index.
   ISBN 0-07-134496-9—
   ISBN 0-07-134495-0 (hc)—
   ISBN 0-07-134497-7 (CD)
   1. Architectural design—Data processing.   2. Computer-aided design.   3. Architectural practice—Management.   I. Title.

NA2728.V665 1999
745'.4'028567—dc21   99-049555

McGraw-Hill

A Division of The **McGraw·Hill** Companies

1 2 3 4 5 6 7 8 9 0 DOC/DOC 9 0 9 4 3 2 1 0 9

P/N 134495-0

Part of 0-07-134496-9

The cover image is The Paul Brown Stadium for the Cincinnati, Bengals Football Team, NBBJ architects.

The cover was designed by Margaret Webster-Shapiro.

The sponsoring editor for this book was Wendy Lochner. The editing supervisor was Caroline Levine and the production supervisor was Sherri Souffrance. The book design and layout was by Clint Wade and Associates.

Printed and bound by R.R. Donnelley & Sons Company.

This book is printed on recycled, acid-free paper containing minimum of 50% recycled, de-inked fiber.

Dedication

This book is dedicated to the design professionals in many disciplines who currently address the challenge to design with digital tools, and to students who are the design professionals of the future.

I thank my colleagues and students, as well as other writers, for all you have shared. In return I offer this book with the hope that it will enable more development of creative thinking skills and design methods to effectively use digital tools, thus helping guide the change to new media.

Design professionals might dedicate the use of digital tools to collaboration and communication that will improve the professions, and protect the environment, while serving the needs of society.

Contents

Preface

mind *That which thinks, perceives, feels, or wills; combining both the conscious and unconscious together as the psyche. The source of thought processes that facilitate the use of computers for artistic expression, design, planning, management, or other problem-solving and issue-resolving applications.*

medium, pl media *The intermediate material for expression. Using computers people work with ideas and information expressed in electronic media.*

new media *Emerging information technology that combines computers, video/television, and telephony.*

If you are one of the many design professionals—people in all types of creative endeavors—who are wondering how to get the most out of emerging design and communication tools, this book is for you. Ever since writing *Mind over Media, Creative Thinking Skills for Electronic Media* (New York: McGraw-Hill, 1993), I have been rethinking and updating approaches to design, responding to emerging information technology, and endeavoring to bring all of this into clearer focus. *Design with Digital Tools* provides many useful methods and strategies for design and communication as well as case studies of some of the best professional practices.

By using digital tools creatively, design professionals, students, and others can develop capabilities to do better work, thus improving the design professions. New media also provide freedoms that many people can enjoy while engaging in creative endeavors, thus helping to fulfill human potential. Knowing how to approach a task and/or visualizing new and better ways to work enable us to discover short cuts that make that work quicker and easier.

To work effectively with any tool requires more than simply having the tool. Effective use demands skill (techniques for using the tool), strategies (methods for working with it effectively), attitude (openness to what the tool may do), and goals (objectives the tool affords). Lack of any of these limits the benefit and increases the frustration of using any tool.

This book provides methods and strategies that can be shared by many design disciplines that use the evolving software and hardware currently available. Thinking skills presented here provide approaches for dealing with the inevitable changes in information technology (IT). Working smarter means saving time and effort, accomplishing more, and finding satisfaction through self-actualization. Using current programs on powerful platforms can increase productivity. Working with the approaches in this book can help achieve improvements that are even more dramatic than those realized from continuous software updates and expensive hardware upgrades.

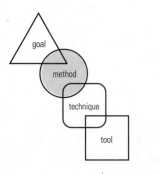

Generic levels

At Cal Poly Pomona, where I am a Professor of Landscape Architecture, my students and I have explored ways we learn to use new tools for design and communication. As a principal in an architectural, landscape architectural and planning firm—CEDG, Inc.—my colleagues and I have been able to find out some of what works (and what does not work) in the context of professional practice. *Design with Digital Tools* has grown out of my teaching and professional practice and builds upon my previous book.

My intent in *Design with Digital Tools* is to engage managers, design professionals, information workers, and students by integrating strategic approaches to change that encompass professional, business, and technical goals. This book focuses on human intellectual capacity and creative potential, emphasizing collaboration.

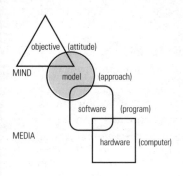

Mind over media

As better tools emerge, there is a growing need to learn ways to use emerging information technology (IT) more effectively. Market research groups—such as Gartner and Standish (in the United States) and OASIG (in Britain)—have found that many investments in IT do not meet their performance goals, are late, over budget, and eventually abandoned. One of the major problems is that a majority of those efforts do not fully address training and skill requirements or properly integrate business and technology objectives. This, and the misconception that new IT tools limit or destroy the human touch in ways that a pen or pencil (also tools) do not, sometimes leads to a backlash in their acceptance.

Design with Digital Tools addresses the human factors—creative thinking skills and organization—that are essential for the effective use of IT. Presented here are case studies of some of the best practices accompanied by a CD-ROM. In addition to providing fine examples of professional work, the CD connects to a professional links library for accessing many information resources on the Internet which are useful to design professionals.

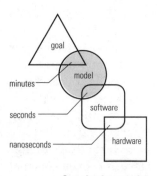

Speed enhancement

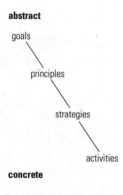

Relating goals to activities

Contrasting discovery and strategy methods

principle *A fundamental truth upon which others are based.*

tool *An object that helps the user to work. This may be a traditional too — such as a pencil — or a digital tool — such as hardware and software for word processing, computer-aided design, animation and many other applications. Used to optimize efficiency where cost-effective.*

This book and its CD relate theory to practice.

Each chapter:

• Distills principles that will help address goals we can relate to.

• Discusses strategies that will help develop creative thinking skills when working with a variety of electronic media to achieve these goals.

• Offers activities that will challenge you to apply these approaches to your own practice.

There are different approaches to learning how to use digital tools, drawing upon different modes of thought. The discovery method is more experiential. The strategy method is more logical. The discovery method proceeds from concrete experience to an abstract understanding. The strategy method starts with abstract goals and principles, which can lead to successful experience.

The activities at the end of each chapter stimulate opportunities for learning and continuous innovation. They provide design teams with methods for determining "what to do" and strategies for "realizing the potential" of available information technology. In this way the entire team can actively consider how to organize work and design tasks to improve effectiveness. Helping to educate and train everyone affected by the technical and organizational changes can enable those involved to gain the maximum benefits from emerging IT. This also provides a basis for evaluating the performance and impact of new media and the creative use of digital tools. Managing IT effectively can build confidence and ensure the willingness to commit sufficient resources for success and fulfillment.

There are twenty-chapters in the book—each relating to a goal addressing needs of design professionals. The book has three parts:

Part I   Methods   "What to Do"

Design teams—including managers, design professionals, and staff—need ways of working effectively. This part of the book offers methods to use new media and digital tools creatively and productively for design communication in professional practice. It can be a helpful text in courses on design methods and new media for design communication, as well as computer courses that go beyond application training.

Part II   Guidelines   "How to Do It"

Good examples demonstrate how progressive design professionals are using new media and applying digital tools in their practices. The case studies examine a range of design practices—writing and graphic design, creating movie special effects, landscape architecture, architecture, and product planning and design. The case studies explore what works in office environments as well as in virtual offices and electronic studios using online collaboration.

Part III Strategies "Realizing the Potential"

Once we invest and commit to using new media and digital tools, it is important to find ways to get the most out of them. This part of the book is devoted to realizing the potential inherent in information technology. It will provide valuable help as your design team addresses change.

CD Organization

The accompanying CD is built around the twenty-one goals (one per chapter) of the book. From the CD file, CD GUIDE.PDF, select the goal you want to explore. Related to each goal is a procedure or guideline summarizing approaches. Related to each procedure is an activity to help you apply the approaches to your own practice. Related to each activity are examples of best practices by other design professionals as well as innovative work by students. Also related to the procedures and activities are a glossary of terms, a bibliography, and online, "The Professional Links Library," which provides a gateway to useful sites on the Internet.

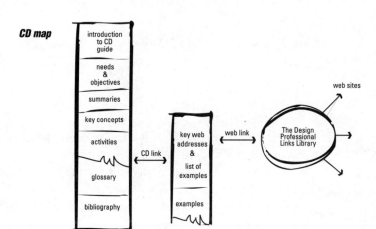

CD map

Tools

For tools to be used creatively, they must be:

- Enabling—address needs and enhance capabilities of design professionals.

- Liberating—nurture creativity without being confining, or inhibiting.

- Simple—not require extensive training or organizational support.

- Unobtrusive—friendly, safe, and unimposing on either the users or the context.

- Lightweight and portable—comfortable to have around the studio, access online, or take into the field.

- Affordable—productivity gains must justify the costs.

Of course, we will continue to use familiar hand tools as well as electronic tools, simply because some traditional tools meet the guidelines listed above and because many people are attached to them. Although this book is about digital tools for new media, I do not advocate abandoning hand tools or traditional media. Pencils and pens are hand tools, and continue to be extremely useful for producing great work in the hands of talented people.

As we carefully select the tools we use and think about how to use them effectively, however, we need to be aware of the exciting array of new tools offering new opportunities. A key to success and fulfillment is how effectively we learn to use whatever tools we select. I am hoping professional organizations and schools can get beyond attaching computers to desks and confining people to cubicles or noisy computer labs. I want us all to find better ways to engage in productive collaboration and enhance our design capability with new media by enabling the creative use of digital tools.

Acknowledgments

Cal Poly Pomona provided me with a sabbatical during which I was able to do much of this writing. As a professor of landscape architecture, I have learned a great deal by teaching and learning to use digital tools with my students. The Department of the Landscape Architecture is a wonderful group that includes other authors.
I feel especially fortunate to have worked closely with John T. Lyle for almost thirty years. I miss him since he passed away last summer, but you will find his influence in some of the ideas presented here. Jared Ikeda, Kyle Brown, and Warren Roberts have been especially helpful on this project.

Claremont Environmental Design Group, Inc., (CEDG) has provided continuous support for this endeavor. As a founding principal of this architecture, landscape architecture and planning firm, I have been able to gain experience applying digital tools in professional practice. Brooks Cavin, the other founding principal of the firm, has been continuously supportive. Erik Peterson, Daniel Cressy, and Jonathan Hartman have helped in many ways.

The extended team in Claremont, where we developed this book and CD, includes Judy Casanova, Sarah Peterson, Clinton Wade, Mary Stoddard, and Elyse Chapman. I am also indebted to the editors and production people at McGraw-Hill in New York, who have delivered this product. There are too many contributors to mention here, but I have attempted to acknowledge them throughout the book.

And finally, I wish to thank my family. Each of you have contributed to the project and expanded my awareness.

While I am grateful for the help of many people, I take responsibility for any shortcomings in what we have been able to present here.

Mark von Wodtke, ASLA
MJvonWodtke@csupomona.edu
July, 1999

About the Author

In the 1960s, when Mark von Wodtke received professional degrees in architecture from RPI and a graduate degree in Landscape Architecture from UC Berkeley, design professionals had very limited access to computers. Mainframes dominated and one needed to use computer terminals to access these expensive tools.

In the early 1970s Mark worked with John T. Lyle and others to establish the graduate program in Landscape Architecture at Cal Poly Pomona and the Laboratory for Experimental Design, which later evolved into the 606 Studio. With funding from the Ford Foundation, John and Mark were able to use geographic information systems running on mainframe computers when doing the lagoon and coastal plain studies for San Diego County.

Later in the 1970s Mark von Wodtke ASLA and Brooks Cavin AIA founded CEDG (The Claremont Environmental Design Group). It was there Mark applied desktop computers that emerged in the early 1980s for specific tasks such as word processing, spread sheet calculations, and drafting.

In 1983 Professor von Wodtke became the Founding Director of the Computer-aided Instruction Lab in the College of Environmental Design at Cal Poly Pomona. The College succeeded in establishing a teaching facility dedicated to distributed computing, which provided design students with access to these tools. As the notion of distributed computing became acceptable, other Colleges on the campus also established their own teaching labs. ITAC (Instructional Technology and Academic Computing) networked these and other labs to mainframe computers on campus and to the Internet.

In the early 1990s Mark von Wodtke wrote a seminal textbook entitled *Mind over Media: Creative Thinking Skills for Electronic Media*, published by McGraw-Hill. This book explored how to link human mental capacity with emerging information technology.

In this new book, *Design with Digital Tools: Using New Media Creatively*, Mark von Wodtke draws upon his experiences as a design professional and educator to provide a guide for creatively using digital tools in the design professions.

I

Methods

"What to Do"

PART I   Methods

Design teams—including managers, design professionals, and staff—
need ways of working effectively. This part of the book offers methods
to use new media and digital tools creatively and productively for
design communication in professional practice. It can be a helpful text
in courses on design methods and new media for design communication, as well as computer courses that go beyond application training.

PART II   Case Studies

Good examples demonstrate how progressive design professionals are
using new media and applying digital tools in their practices. The case
studies examine a range of design practices—writing and graphic
design, creating multimedia and special effects, landscape architecture,
architecture, and planning. The case studies explore what works in
office environments as well as in virtual offices and electronic studios
using online collaboration.

PART III   Strategies

Once we invest and commit to using new media and digital tools, it is
important to find ways to get the most out of them. This part of the
book is devoted to realizing the potential inherent in information technology. It will provide valuable help as your design team addresses
change.

APPENDICES

1

1

Learn to Navigate and Work in Information Environments

New media help us gather information quickly, and add value to it by developing and communicating designs. The keys to doing this are learning to interact and navigate online.

Our mindset—attitude—has a major influence on what we can do with new media. A positive attitude maximizes the possibilities and minimizes the problems. A passion for new media makes using them exciting. Persistence also pays off. The methods, strategies, and activities in this book are intended to provide good experiences, which will nurture a positive mindset.

Having so much information at our fingertips, the potential to express ideas in multimedia splendor, and the capacity to communicate online can be mesmerizing. The possibilities of losing information, wasting time and money, or embarrassing ourselves with new media can be terrifying. Try not to let either fascination or fear detract from what you need to do. Learning to balance possibilities with practicality is crucial to the creative use of digital tools.

Mindset

Using new media can be liberating. The Internet provides new worlds to explore. Multimedia helps interactively compile information and ideas in many different ways. Electronic communications enable collaboration online—anytime, anywhere.

Enhancing Enjoyment

There is a flow channel–that path between anxiety and boredom—that provides satisfaction when using digital tools. Multimedia computing can have all the elements that enhance enjoyment. As Mihaly Csikszentmihaly points out in his book *Flow: The Psychology of Optimal Experience,* these include a challenge requiring skills, a chance of completion, the opportunity to concentrate, the merging of action and awareness, clear goals, immediate feedback, deep involvement transcending distractions and the awareness of time, a sense of control over actions, absorption of self, and expansion of self through experience.

Caution! New media can be addictive. It is not unusual to lose track of time when absorbed in electronic media. Avoid the temptation to use them everywhere—all the time. Balancing mental activities, physical exertion, and rest is essential for a healthy life.

The Flow Channel adapted from "Flow: The Psychology of Optimal Experience" by Csikszentmihalyi

Navigating

Computer operating systems provide access to files. We can picture directories (or folders) of computer files as tree diagrams, showing subdirectories as branches or roots. Metaphors help us navigate. The Macintosh system uses desktop metaphors; Windows for MS-DOS and X-Windows for Unix use window metaphors enabling us to peer into several different places simultaneously. Graphic user interfaces with pull-down menus help us select commands by pointing and clicking with a mouse or other pointing device. It is also possible to use voice commands.

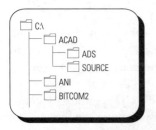

Disk directory

Professional offices, or workgroups, should have file-naming conventions. This makes it easier for people working together to place and find files in their information environment. The convention should work across platforms—between Mac, Windows, and Unix operating systems—which may limit file names to eight characters with a three-character extension. The standard should include a consistent way of naming files using references to job numbers, description of contents, and version or author. The software automatically records the date, file size, and type of document. For example, the first three characters could be the office job number, the next three characters a content code, and the last two characters the version or the initials of the author. Most software adds the three-character extension indicating the file format. With this eight-character file-naming convention, it is easy to use an operating system to sort by job number, file contents, version or author, or file type. Longer file names can be more descriptive, but may not sort as easily or may become truncated when moved between operating systems.

File-naming convention

On the CD that accompanies this book are examples of the file-naming system used for creating it, as well as the file-naming system and office policies for managing our information environment at CEDG, Inc.—the architectural, landscape architectural, and planning firm—where I am a founding principal.

If files are clearly organized in directories, or folders, on a local area network (LAN), everyone using this information environment can learn where to find files and applications. This is important, even when using a "sneaker net"—moving information on disk between several computers. Naming files consistently, and knowing where they are, will help avoid losing information. It also makes it easier to keep track of the latest version of files and applications.

compact disk-read only memory (CD-ROM) *A laser disk for storing digital information. Typically has the capacity of 650 megs.*

Internet *The public online environment that includes e-mail and other information services including the World Wide Web.*

Intranet *An information environment established by an organization for internal use. Can be navigated using same tools used on the Internet.*

World Wide Web *The information environment accessible through Web browsers such as Netscape Navigator or MS Internet Explorer.*

Uniform Resource Locator (URL) *The address of a site on the World Wide Web.*

Network software provides tools for navigating a LAN. When connecting portable computers to a network sharing files and peripherals, be careful about uploading and downloading files and consistently update the files on the LAN. Use temporary storage directories, such as "My Briefcase" in Microsoft (MS) Windows. Other operating systems and utilities also provide effective ways of comparing versions of files when transferring them from notebook or palm computers.

Integrate a backup system into the LAN so that everyone using the system can store files in a consistent way. This may involve saving in several places on a network or copying to a storage device with removable media. Organize files on your backup system to reflect your information environment so that people can navigate it easily, and get everyone on your team to back up consistently. Burning a CD periodically—perhaps once a week—and keeping the CD at another location (like a safety deposit box) can protect your information in the event of disasters. Archiving project files to CDs is a compact way of storing information and may be less prone to degradation over time than magnetic media. Archiving files on CD-ROM can free space on hard drives and still provide quick access to the information. Of course, hard copy of important documents—on archive quality paper—is time-tested and one needs no electronic device (other than maybe a light bulb) to read it.

Many organizations are using Intranets that people can access with the same software they use to browse the Internet. Netscape Communicator and MS Internet Explorer are programs for navigating the World Wide Web. Both use a page metaphor enabling one to go from a home page to other places on the web and return at any time. Service providers offer home pages—which serve as gateways, or portals, to the web. Links enable movement within documents or to other web sites. Favorite places can be "book marked." Browsing is done clicking on links or by moving forward and back. Search engines—such as Yahoo, MetaCrawler and others—search by author, topic or key words to find URLs. Navigation software also guides us (the clients) to servers with interesting places on the web and provides access to the Internet so we can use e-mail.

agents *Bots(robots) or droids (androids) that help gather specific information.*

Agents—such as bots or droids—can help gather specific information we are interested in. Chatterbots, such as "Ask Jeeves", enable the user to ask questions and get responses through natural language processing.

hypertext markup language (HTML)
Defines the layout of text and graphics on web pages.

eXtensible markup language (XTML)
Describes the content of a web site using standard tags for links.

virtual reality markup language (VRML) *A file format for three-dimensional web sites.*

upload *To transfer digital information from a local computer to a remote server (often a mainframe computer).*

download *To transfer digital information from a server (or mainframe computer) to a local computer.*

File transfer protocol (FTP) *The software code for uploading and downloading digital files.*

applet *A small application program that can be downloaded from the Web.*

Java *A cross-platform programming language for developing applets.*

digital versatile disk (DVD) sometimes called digital video disk
A high-capacity laser disk for storing digital information. A single-sided DVD can hold up to 4.7 GB on one layer and 8.5 GB on two layers. A two-sided, two-layer DVD has a total capacity of 17 GB.

mindmap *A diagram of associations showing how you link key words.*

The CD-ROM that accompanies this book connects to a Design Professional Links Library which provides current URL addresses useful to design professionals. Brief annotations describe what each site has to offer. The intent of this "Links Library" is to enable professionals and students of design to quickly find and get to web sites with useful digital information and tools. There you will find links to useful information sites selected by design professionals.

Applets are small programs, usually written in Java, which we can download and use in our own computer. Applets can expand functionality by performing specific tasks.

Currently, Web pages use HTML and XML to define links and content. As development of the World Wide Web evolves, more three-dimensional virtual environments will emerge using virtual reality markup language (VRML) and other three-dimensional formats for video games, entertainment, and education. A spatial metaphor, such as a walking tour, can be can be particularly useful for the design communication of three-dimensional objects and environments. Because people are as familiar with navigating in space as they are with paging through books, it can provide an intuitive interface for organizing information. Hardware and navigation tools could include 3D headsets with surround sound.

Audio and video tape recordings (DAT and VHS tapes) are linear—they follow a time sequence. To navigate one needs to remember the sequence of events. A counter serves as a reference, but often certain sounds or scenes indicate where we are on the tape. To find something, fast-forward or reverse through the tape. Obviously, this medium is well suited for listening to music or viewing movies—sequences we want to experience from beginning to end. Digital audio and video editing provide the tools to change that sequence.

Compact discs (CDs) and digital versatile discs—sometimes called digital video discs—(DVDs) provide access to large amounts of information nonlinearly, which we can randomly select using a browser. Navigating topics in non-linear ways helps explore information more interactively. Hyperlinks build upon associations that should reflect our thinking patterns—sometimes referred to as mindmaps. The CD accompanying this book links goals to procedures that address those goals, and to activities to help apply those procedures to one's own design projects. The CD includes the Adobe Acrobat viewer for navigation.

Mapping Information Environments

cyberspace *Media space connected to the human brain, enabling people to experience this information environment interactively.*

media space *The information environment connecting real and imaginary places and the people and objects within them. The environment in which people can use representations to work with artificial reality.*

artificial reality *A model or representation of reality. People can develop this model using information in media space.*

virtual reality *A simulation using information to provide, in effect, realistic experiences. People can create this simulation by using computer-generated images in media space.*

mindscape *The inner world of your own mind, involving both conscious and subconscious levels.*

e-business (electronic business) *Encompasses e-commerce, and includes applications to help business run more efficiently. It also includes more internal applications for linking employees together and helping employees work more productively. E-business also involves publishing and accessing information.*

e-commerce (electronic commerce) *Involves buying and selling (on the Internet) and all the processes that support buying and selling, such as advertising, marketing, customer support, and credit-card activities.*

e-gov (electronic government) *Providing governmental services and information online.*

e-mail (electronic mail) *Messages transmitted by computer to addresses on a server. These messages can be picked up anytime, from anywhere, using a computer to access the address.*

We can, in effect, create and experience information environments with computers and interactive video. William Gibson, a science fiction writer, refers to these environments as *cyberspace*. The popular press often uses this term today. In 1985, researchers in the System Concepts Laboratory at Xerox PARC coined the term *media space* and explored new ways of working electronically. Michael Sorkin wrote about the *Electronic City* and its implications for society—some of which are emerging today on the World Wide Web. Jaron Lanier coined the term *virtual reality* to refer to electronic environments that we can enter and experience—now provided by popular video games. Designers can create digital models as prototypes for implementation in reality and for special effects in movies. We may also design web sites for people to visit. Even in organizing our own information environments, we shape our cyberspace.

Learning to navigate in cyberspace is an important notion that requires developing a sense of place when working in an information environment. Cyberspace can be an extension of our work environment for design and other collaborative endeavors, just as e-mail is an extension of communication; e-commerce, an extension of e-business; and e-gov, web-based governmental services. Media space connects real and imaginary places and the people and objects within them. Information environments (which many see as the next frontier) are growing rapidly and becoming more vivid as the bandwidth of communications increases.

Emerging information environments are having dramatic effects on organizations and the workplace. Because moving ideas becomes easier and faster than transporting people, electronic communication is changing the way we collaborate and commute. Information can be accessed anywhere, anytime. Information workers, collaborating in the freedom and flexibility of media space, are changing work patterns. New opportunities are emerging for education and life-long learning. It remains to be seen what we will do with these freedoms and opportunities.

In his book, *Image of the City*, Kevin Lynch observed that people develop a mental image of a city. By identifying landmarks we develop a sense of place so that we can find our way. The maps people draw of their image of a city depend upon what they understand and have access to. (Children might just know the neighborhood where they live, while a taxi driver will know most districts of a city and the routes to get there.) In the work environment we also have a mental image of where we put things in our file cabinet, desk, briefcase, and notebook. A clear mental image of our media space is just as necessary.

Visualizing media space

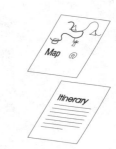

Visualizing physical space

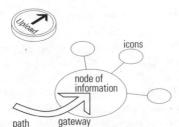

Cognitive map of media space

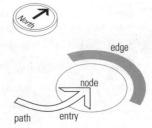

Cognitive map of physical space

directory A guide. The directory of a computer disk will provide a guide to its contents.

subdirectory A folder contained within a directory. A computer disk can be divided into many subdirectories.

cognitive map A visual representation of what you know. A cognitive map can show spatial or conceptual relationships.

portal Gateway to nodes or sites on the World Wide Web.

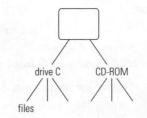

Branching diagram of disk space

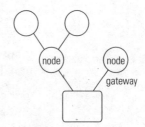

Branching diagram of media space

This is especially important if we use shared information environments and need to let others know what we have named a file and where we have put it. Working with the information context that is available and shaping it to meet our needs, we become the architects of our own media space.

A branching diagram shows *directories* (folders) and *subdirectories* where we place files. We can diagram portals (gateways to nodes or web sites) we have access to online and identify the paths we use to get there. These *cognitive maps* help us develop a mental picture of where we are, what we can access, and how to find our way. We can provide directions and draw maps showing others on our design team how to find files and find their way in an information environment.

As in our physical environment, we also develop domains of community and privacy. We consider the World Wide Web to be public, our local area network to be semi-public, and what is in our e-mailbox or desktop to be private. Some organizations go to great extents to build firewalls protecting information, and prohibiting access by competitors or hackers who try to penetrate information environments. Organizations, like banks, may even use private Wide Area Networks (WANs) to protect financial information. Using the Internet, Virtual Private Networks (VPNs) also provide private pathways within a busy public network. VPNs are less expensive than WANs and offer greater flexibility.

It is relatively easy to establish electronic information environments—web sites with domains that can grow to meet evolving needs or shut down once a project is over. It is much more difficult to change our physical environment.

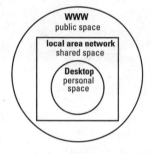

Personal/shared/public space

hub *An information site that provides access to the other sites.*

Today, information resources are becoming more dispersed as well as more interrelated. The computer, television, and telephone have merged. Networks and modems can link information stored in many different computers. Now, when we look at a computer monitor, we can have a window into a rather extensive information environment. We can access information services that provide access to a wealth of current information online, as well as library card catalogs listing archives of printed matter. Local area networks and modems can help to more effectively share the information that we have at our workstation. Local hubs enable us to communicate all over the world.

Our cyberspace expands as we learn new applications, use telecommunications, and explore the World Wide Web. We can sketch *maps of media space*. By developing cognitive maps using a computer graphic program, they become easy to update and share with others in our workgroup so we all know how to find information, applications, and services online. By developing clear cognitive maps of our information environment, we can work online more successfully. With that comes a sense of confidence in using electronic media.

There is a difference between navigating and authoring—much like the difference between being a tourist and a resident. When you are navigating, you experience cyberspace and need to find your way within the information context that is available. When you are authoring, you need to shape media space so that you, and others, can work with it creatively. As an author, you can, in effect, become the architect of your own media space—you can establish a framework and begin to organize information, adding your own insights.

Map of media space at CEDG, Inc.

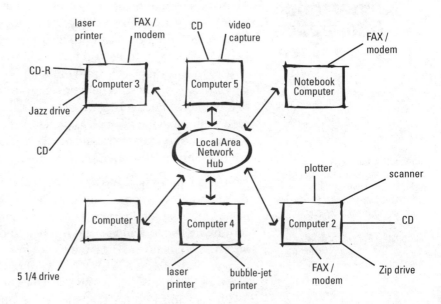

information work *Thinking work
involving the transformation of information by human brains or computer
programs.*

digital nervous system *Using information technology to extend our
capabilities to perceive and respond.*

Michael Dertouzos recognizes information as-a-verb (and not a static
archive) in his book *WHAT WILL BE: How the New World of Information
Will Change Our Lives.* In the book *Business @ the Speed of Thought,* Bill
Gates describes how we can use "digital nervous systems" to link
people with business, government, and education. Web workstyles
and lifestyles are emerging that actively use digital information in
real-time.

Diagramming Information Flow

information flow diagram
A graphic representation that helps users visualize how to transfer information into a digital format that they can work with interactively. Especially helpful when integrating information from many different sources and transforming it into a range of products.

	source info		final product
text			
graphics			
images			
video			
sound			

info types

Format for information flow diagram

Think of information as being embodied in objects—programs, documents, models, and graphic representations. By representing information objects with boxes and linking them with lines, we can graphically diagram how to transfer information from one object to another—or how to transform objects from one stage of a project to another. These *information flow diagrams* help develop strategies for transferring information more effectively. Such diagrams are also helpful when working with traditional media such as paper or film, but become more so when working with electronic media, because there are so many different pathways for moving information.

I like to set up a chart showing the source information in columns on the left and final products in columns on the right, with different channels of information in rows. This helps examine how to transfer and coordinate text, drawings, and images. Another approach is to keep information related to different topics in separate rows. For example, when working with a landscape, keeping site information in one row and user information in another clarifies how to merge this information to develop a design that addresses both site and user needs.

Simple information flow diagrams drawn with pencil on paper work fine to share around a table. Drawing them, using a computer graphic program, provides a digital file that is easier to update and share online. These diagrams can help develop strategies for working with any medium—text, graphics, images, even video or sound. We can start with the information we have and diagram strategies for what we can do with it, or envision what the final product will be and work backward. Using both approaches together we develop a more complete picture of how to transfer information.

Visualizing strategies for transferring information helps us avoid gathering information that is not needed, and discover where there are dead ends or missing pieces of information. By diagramming information flow, we can examine transfers at each stage of the project and determine the most effective way to integrate the necessary information into the final product.

Information flow diagram for plan rendering

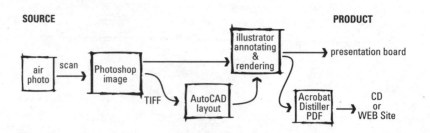

10

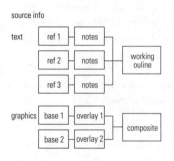

Start with source information

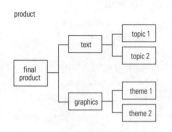

Start with the final product

Diagramming Data Flow

As we gather information, we get a better idea of how to work with it in the cyberspace we are using. Often, information we thought we could acquire isn't available. A clear picture of information flow enables us to work with surrogate information or proceed with assumptions that we can verify later. Sometimes we will discover new information that is useful to the project. We may also change or refine the project goals as we proceed. Obviously the information flow diagram evolves as we find out more about the project, and it may also evolve as we learn more about the media we are using.

We can use maps of cyberspace and information flow diagrams much like we use maps of a transit system and itineraries. They help us to develop clear strategies for attaining a predetermined goal or to simply establish a general direction and sort out what we discover as we proceed. Try not to simply replicate traditional approaches when using new media. Just as the route we take when flying is different from the one we take when driving a car, the process we use with each vehicle for communication is different. We should make sure our information flow diagram fits the tools we are using. Electronic media offer many opportunities for productivity gains. But using digital tools to simply replicate what we are doing by hand is often not worthwhile. We can enhance our capabilities by using flow diagrams to think through better ways to apply digital tools that are emerging.

Diagramming data flow can help us figure out the best ways to move files from one program or computer platform to another, thus avoiding the risks of not being able to make the transfer and having to start over. Boxes can represent the files in different formats. Arrows can represent the transfers involved. The framework we work with can relate to our media space. For example, we can visually show how to upload and download files. Once we diagram a strategy, it is important to test it to make sure the data transfers work. Developing "living files" that evolve into our product can save considerable time and effort. Some applications now provide for *dynamic data exchanges*. Changing information in one file, such as a spreadsheet, will change information in another file, such as a word processing document. GIS and some CADD programs also dynamically link objects (such as polygons) with attributes. Changing the object updates the attributes.

Maps of media space and flow diagrams help us find our way— especially when we venture into information environments and applications we haven't used before. Even when working with familiar applications, these visualization techniques can help us find shortcuts. Each person in a workgroup should know how to navigate, where to put files, and how to transfer them efficiently.

Summary

Methods for Navigating and Working in Information Environments

1. **Become familiar with the tools that operating systems, browsers, and computer applications provide.** Learn to work with digital files and use search engines to access information online. Learn to navigate and work with useful applications or collaborate with those who can.

2. **Use shared information environments where you can work collaboratively with your team.** Developed cognitive maps to visualize these interactive information environments

3. **Establish file-naming conventions.** Organize files in directories or folders to quickly find them again.

4. **Establish backup procedures and use them consistently.** Data flow diagrams can help workgroups understand these procedures, and make it easier for everyone to back up information they are responsible for.

5. **Develop clear strategies for using information and transferring files from one application to another.** Information and data flow diagrams can help workgroups develop effective strategies for working with digital files. We can move files on our desktop and local area network, send files attached to e-mail, or download or upload files to servers using file transfer protocols.

6. **Learn to manage files.** Keep track of the latest versions, deleting useless files and closing applications to make sure you have sufficient disk space and random access memory (RAM) for the tasks at hand.

7. **Maintain a directory of e-mail addresses.** Collaborate online with other people by exchanging correspondence and attached files.

8. **Expand your information environment.** Save bookmarks of web sites that are useful so that you can quickly access them again online. Maintain a collection of CDs that have applications and information you can access from CD-ROM, perhaps including an archive of your past professional work.

Activities

Navigate and Work in Information Environments

Design professionals need to communicate effectively and access a great deal of information— a growing amount of which is becoming available on the Internet. The challenge is to develop collaborative teams that can work together online, quickly gathering and disseminating information. Clients, user groups, and constituencies can also become part of this online collaboration.

Select a project to use as a case study for making better use of the information technology you have available. Do a project where communication is especially important, the consulting team is in different locations, digital base information exists, and the client is already computer literate.

Begin by carefully considering what you can do with the tools you already have access to. If necessary, invest in hardware and software as well as training to be able to communicate effectively online. You may want to obtain support from those who have information and technology that will be useful for the project.

Assemble a team that has the necessary capabilities. Include the knowledge base to carry out professional roles needed for the project as well as communication skills using digital tools for new media. The team should develop core competencies— working with operating systems, Internet browsers, e-mail, and file transfer protocols—to be able to collaborate online. Use this book to develop a common language of designing with new media, helping to team experienced professionals with computer-capable staff. Include outside consultants with professional knowledge, digital tools, and technical savvy. Consider using

service bureaus that provide online information services. Get technical support from hardware and software vendors to help your team reach new levels of performance using information technology.

Meet together—either in an office or online—to discuss the project goals and scope of work, assigning responsibilities as you normally would in an organizational meeting. In addition, go through the methods outlined in this chapter. Develop clear cognitive maps of information environments you have access to; set up a shared information environment for working collaboratively using digital tools; establish file-naming conventions and backup procedures, diagram clear strategies for using information and transferring and managing files. Make sure you exchange e-mail addresses and URLs of key sites so that the team can continue to collaborate online addressing the project in hand.

Build the team's capabilities to effectively collaborate online. Learn together and develop working relationships to build upon for the next project. Consider creating a web site to provide more opportunities for online collaboration. This would enable your team to disseminate multimedia project information to the growing number of people using the World Wide Web.

2

Use New Media to Gather Information, Perceptions, and Ideas

Digital tools can help us capture what we see and hear. This can stimulate our imagination and become a source of innovation.

Creativity emerges from many areas of our minds. We can draw upon our whole brain using new media. *Multimedia* opens new and exciting possibilities for relating computing, telecommunications, publishing, video, and TV. Electronic media also can be integrated with manufacturing and construction in the creation of real objects.

We now are able to create personal portals to the Web, customizing this site to provide current information we care about—such as the local weather, or information related to projects we are working on. Personal portals, searching the web, can help us compile a digital design journal to which we can add our field observations and other information we compile from different sources. The real challenge is to work creatively with all this information.

Multimedia

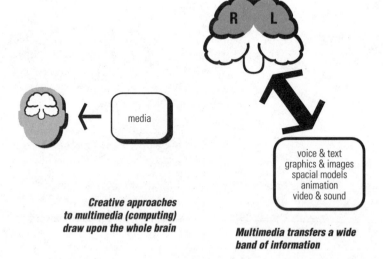

media

Creative approaches to multimedia (computing) draw upon the whole brain

R L

voice & text
graphics & images
spacial models
animation
video & sound

Multimedia transfers a wide band of information

Interacting

Multimedia changes the way people think when using computers. Conventional computing dealt primarily with numbers and text. Today, telephones, computers and graphic workstations, as well as interactive video devices, provide access to a broad *bandwidth* of information that engages many of our senses electronically. Multimedia provides new opportunities for learning, creating and communicating.

Interacting involves viewing and doing. This is different from simply watching. We can work interactively with a wide range of digital media right on our desktop, linking mind and media.

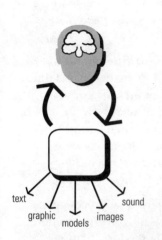

Interacting—alternately viewing and doing what you visualize

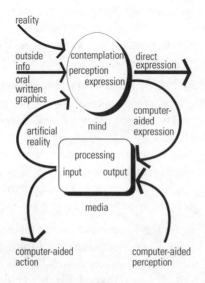

Coupling mind & media

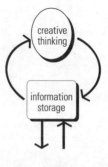

Couple interactive browsers, viewers and applets

multimedia *Integrating more than one medium. Computer systems can enable the integration of electronic media combining text, graphics, animation, spatial modeling, imaging, video, and sound.*

bandwidth *A range. Multimedia encompass a wide bandwidth.*

Communication involves moving ideas and information. To be able to access the information highway, we need a good computer and the ability to use it. (Or have computer operators be our chauffeur.) In addition, we also need a mental model of where we are going—an objective and the ability to find our way. This involves both mind and media. To work interactively with media using computers, we need to develop thinking skills that help us navigate and work interactively.

Just as it is important to have access to multimodal transportation systems, it is also important to have access to multimedia communication systems. In some situations it makes sense to talk on the phone, or write a handwritten note, or make sketches we can fax. Using telecommunications, we can access databases, move documents, even interact with video. Both transportation and communication provide a sense of freedom. With transportation we have the potential to go places and move goods. With communication we have the chance to explore ideas and transfer information. Working with a computer, we can compose a message and send it by e-mail or other forms of telecommunications. Other people can read, respond at their own pace, and then forward the message (and their comments) to others in other places.

The trick to being able to use electronic media interactively is learning how to effectively transfer ideas and information into cyberspace. A common pitfall is using the wrong tools. We especially need to develop good approaches for using the tools we have access to.

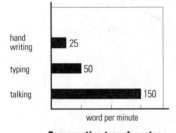

Comparative transfer rates

We can transfer information and ideas to media at different rates, depending upon the tools we use. We can capture images in real time on video and audio recordings faster than our mind can assimilate them. (No doubt you have seen an instant replay and know the experience of discovering much more of what happened than you were aware of when you experienced it the first time.) Digital cameras instantly capture images. Translating mental images into drawings takes longer. Translating mental images into drawings takes longer.

Translating images into words is a comparatively slow process since we think visually at an equivalent of thousands of words per minute. For this reason, we should do initial transfers in the form of quick notes or idea sketches that enable us to record our thoughts as they are occurring. The next quickest mode of expressing words is through talking. Average speech takes place at 150 words per minute. Voice recognition is now available for computers. Shorthand can almost keep up with verbal expression; typing is somewhat slower—between 30 and 90 words per minute or about one-third of the rate at which people talk. Handwriting is slower yet—typically between 15 and 30 words per minute. This is, on average, about one-half the rate at which people type or about one-sixth the rate at which people talk.

Unless our initial transfers are quick, we risk not getting our ideas down. Using key words and outlines or idea sketches and diagrams compresses information. When working verbally, we can capture key words. We can make outlines using a word processor or on a notepad using a pencil. Audio and video recording enables real time capture of events. Speaking spontaneously or from notes, we can use voice recognition software or dictate on tape and then transcribe it. We can brainstorm or expand an outline using a word processor. Similarly, working graphically, we can make idea sketches on paper and then scan them into a digital format. We can develop idea sketches using graphics or CADD programs. Once we have a composition and some symbols in the computer, we can rapidly manipulate them electronically. Drafting by hand, like writing by hand, is slow compared with the other methods. Handwriting and sketching are very effective means for mental note taking, but we can go beyond these basic tools. Just as keyboard skills enable us to express written compositions digitally, mouse or stylus and other digital tools enable us to express graphic ideas using electronic media.

Transferring ideas to new media makes it possible to use digital tools to develop them creatively. Many forms of electronic media—like word processing documents—are extremely flexible. We can add information and make changes easily. As more and more people work electronically, the process of sharing information and communicating over long distances in less time becomes easier.

Using the Internet, we can subscribe to list servers that can compile information or put us in touch with other people who share our interests. On the World Wide Web there are search engines that help us find information. Many educational institutions, governmental services, commercial enterprises, and service organizations have web sites. It is also possible to access databases, encyclopedias, maps, etc. Many libraries also have computerized card catalogs we can access online.

Beyond selecting and receiving information, telecommunications enable us to act on the information we receive. We can download computer programs and files, shop and conduct financial transactions, and plan trips—booking accommodations and making airline reservations. We can collaborate with other design professionals, visit chat groups for advice, and participate in continuing education online.

Computers have the potential for providing a much broader range of stimuli when they include video. Fiber optics and other improvements in telecommunications, along with data compression, are making it possible to transmit live video images. Personal computers and TV are just now integrating in ways that will empower individuals to browse electronic media more effectively.

Interactive video provides even more possibilities. Viewing and transmitting video setups, participants can communicate in much the same way they use a telephone, although with a dramatically broader bandwidth of information. Commercial video conference centers enable us to confer with people at centers in another city. Some schools are using interactive video, enabling specialized instruction at locations offsite and providing the audience a chance to participate.

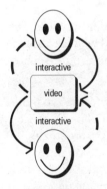

Interactive video

interactive video
Video that is both sent and received. Enables users to interact in real time involving both audio and video transmitted electronically.

17

Computers provide a multimedia vehicle for *multidisciplinary* and *multidirectional* collaboration. As the bandwidth of electronic media widens, we can learn to work creatively in cyberspace—more like an artist than a tourist—studying and interacting with what is there. Disciplines can collaborate in this shared media space.

The possibilities continue to expand. Electron microscopes, remote sensing equipment in satellites, and other devices help transcend human physical limitations. Infrared, x-rays, ultrasound, and magnetic resonant images enable us to experience what we cannot normally see with the naked eye. Interactively we experience virtual realities. We can explore digital models of new landscapes or buildings before they are built. These realistic experiences enable us to draw extensively upon our mental capacity. Just as we can learn to make perceptive observations when dealing with reality, we can also learn to work perceptively with representations or models of reality.

Assimilating Information

Assimilating information is like digesting food. Sometimes we can suffer from overeating (information overload). On the other hand, we can hunger for good information and sensory stimulation. We need an appropriate information flow. Multisensory stimuli provide a balanced diet; they nourish the whole mind. We can browse information, satisfying our interests like we can sample foods. We consume kernels of information (often in sound bites or in visual fragments). Pursuing trivia is much like nibbling. Studying should be like eating a well-balanced meal. Absorbing information nurtures our psyches as digesting food nurtures our bodies.

Recognizing that we can't deal with all the information that is available, we need to find ways to select and assimilate what interests us. There are both verbal and visual approaches to organizing information. Alphabetical lists are primarily verbal, although we can often visualize what is at the top and the bottom of the list. We record and calculate time numerically, but picture this by means of a watch face or time line. We can describe locations verbally, but diagrams or maps provide a clearer picture. We can work with a continuum and magnitudes numerically, but graphs and charts provide ways to visualize these relationships quickly. Although we can describe categories verbally as lists or outlines, branching structures provide ways to diagram categories. Displaying complex information visually increases comprehension.

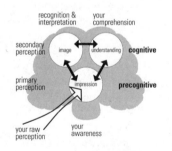

Perceptual process

Hypermedia (like we find on the World Wide Web) provide ways of organizing information by association. Links provide threads for the

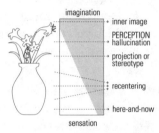

imagination

→ inner image
PERCEPTION
→ hallucination

→ projection or stereotype

↕ recentering

→ here-and-now

sensation

Perceptual filter

reader to explore by selecting subtopics and probing them further. Hypermedia make it possible to package information in ways that users can explore more interactively.

Visual thinking skills develop more awareness and comprehension of what we perceive. This enables us to work with representations—drawings, diagrams, models, animation, and videos—and express ourselves better using multimedia.

Perception (a blending of sensation and imagination) is filtered by many factors, as illustrated in the diagram to the left. The visualization techniques that follow help develop perception.

Gestalt

The overall sense of comprehension that transcends physical reality is sometimes called *gestalt*. Gestalt manifests itself in many ways. A good painting is more than just pigment on canvas; music is more than just a series of the notes; a personality is more than just a set of traits. Gestalt psychology provides ways of appreciating and nurturing the more intuitive and integrative aspects of perception. It also applies to how we perceive the artificiality of an information environment.

Levels of Perception

Primary perceptions (impressions) sometimes register only in the subconscious. *Secondary perceptions* (images and understandings) register more in cognitive modes of thought. Each of our senses of perception has primary and secondary levels. From this, we derive meaning.

Classic ways of refining awareness and enhancing comprehension include discussion, writing, drawing, modeling, and composing. Digital tools enable us to do these things interactively. We can refine observations, share them, and transfer them to other applications. For example, we can take notes on what we observe; then, using a word processor, write about it. We can integrate what we write into a letter or a report. We can follow similar approaches using computer graphics programs.

The arts and design disciplines emphasize imaging as a basis for awareness. Science and engineering disciplines stress cognitive understanding as a basis for comprehension. An architect and an engineer can look at the same thing but perceive it quite differently. This may be the source of many communication problems but it also provides the basis for effective collaboration. Some people seek awareness; they derive images that provide them with a sense of what they experience. Others seek comprehension—deriving understanding from what they see, hear, or read. Ideally, we develop both our awareness and comprehension.

Modes of Thought

Our minds have many facets. People often refer to what they feel in their heart as opposed to what they understand in their head. Right-brain thinking tends to involve the whole body, which people associate with the heart. Left-brain thinking tends to be more cerebral—related to the head.

In her books, *Drawing on the Right Side of the Brain* and *Drawing on the Artist Within*, Betty Edwards distinguishes between the *R-mode*, which deals with visual thinking, and the *L-mode*, which deals with verbal thinking. The R-mode is more holistic, using inductive thinking to deal with perception and comprehension of relationships; the L-mode is more segmental, using deductive thinking to deal analytically with sequences and details. Inductive thinking tends to be intuitive; deductive thinking is more logical. The R-mode relates to what we perceive; the L-mode relates to what we understand. We can shift between these modes of thought.

The same dichotomy exists between modes of thought characterized by *in* words (*induce, innovate, and invent*) and those characterized by *de* words (*deduce, derive, and develop*). The *in* words characterize inductive thinking; the *de* words characterize deductive thinking. Both are important modes of creativity. Inductive thinking is more characteristic of the arts. Deductive thinking is more characteristic of the sciences. Both art and science, however, involve inductive and deductive thinking. Often, intuitive leaps can provide new breakthroughs for understanding. New levels of understanding, in turn, provide platforms from which we can take further intuitive leaps. Although initially computers were considered highly deductive tools, the interactive nature of electronic media lends itself to shifting between inductive and deductive thinking.

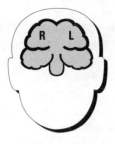

L-mode -- deductive thought processes based on logic

R-mode -- inductive thought processes based on intuition

Right & left brain

Right Brain "R" Mode	Left Brain "L" Mode
impressions	understandings
holistic	segmental

Inductive thinking "in" words	Deductive thinking "de" words
induce	deduce
innovate	derive
invent	develop
inspire	describe
intuit	define

Internal Transfers

Often, people have the tendency to favor one mode of thought over another. This may be like the tendency to favor one hand over another. Most people seem to be right-handed but left-brained.

We can learn to use both hands and our whole brain. Just as each hand can perform important functions when we do physical tasks, each mode of thought is crucial to creative thinking. We need to learn how to make internal transfers from L-mode to R-mode thinking. For example, we can get in the habit of drawing or diagramming a whole image of what we are dealing with, while also examining the parts. Robert McKim, in his book *Experiences in Visual Thinking*, calls using both modes of thought ambidextrous thinking.

proposing

disposing

proposing

disposing

Alternating current of design adapted from "Design for Human Ecosystems" by Lyle

Creative thinking involves alternating between right-brain and left-brain modes of thought and between internalized thinking—reflection—and externalized thinking—expression—when proposing and disposing of ideas. In his book *Design for Human Ecosystems,* my colleague and friend John Lyle called this "the alternating current of the creative process." Although we shift between these modes of thought naturally—almost like breathing in and breathing out—we may get stuck in one mode or another. Returning to this alternating current will help breathe life into our creative approach.

Different modes of thought also relate to different modes of learning. R-mode learning is more experiential and is characterized by statements such as "I have a feeling for that" or "Let me get into that." This learning experience is more holistic. L-mode learning is more rational and is typified by statements such as "I understand that part" or "Let me read about it." The learning experience for L-mode thinking is more segmental, like putting together pieces of a puzzle.

try

prepare

Alternating approaches to learning

The wholeness principle, articulated by Stèphano Sabetti, integrates internal transfers of energy. This synthesis relates to the concept of soul—an ancient idea—manifested across many cultures. The Life Energy Process® developed by Sabetti helps access, explore, and express our whole being. There is tremendous potential for sharing this energy in ways that resonate with others, not only through direct personal interaction, but also through caring collaboration using the World Wide Web—an emerging energy field that transcends our individual beings.

Witnessing

Contemplating

Meditating

witness *To personally experience. Can involve centering on where you are, to become fully aware of your senses.*

contemplate *contemplate To access different modes of conscious thought. To reflect upon what we experience and understand.*

meditate *A way of thinking that enables people to access altered states of consciousness. To project into unconscious modes of thought.*

Creative thinking also involves moving between the conscious and subconscious levels of our mind. We can do this through witnessing, contemplation, and transcendental meditation.

Witnessing involves centering on where we are, becoming fully aware of our senses. We witness when we sit in a beautiful landscape and absorb all of what we perceive—the subtleties of sunlight, the sound of birds, the fragrance of the air. Witnessing also can be useful when interacting with multimedia.

Contemplating enables us to access different conscious modes of thought. In this way we can reflect upon what we experience and understand. For example, we might contemplate an event in our life by replaying it in our mind's eye and relating what we understand and how we feel about the experience.

Meditating enables us to access altered states of consciousness. We can do this through active imagination and dream work. For example, we can take fantasy flights and discover things we were not even conscious of, thus projecting into unconscious modes of thought.

Expression

The mind provides the ideas—inspirations, brainstorms, hunches, feelings, and connections or associations. The progression we go through in transferring ideas from our mind to any medium usually starts internally with some sort of insight or muse. Initially, we may not even express our ideas. We may simply reflect on them. Given the opportunity, we can express our muse with simple gestures—a quick sketch or jotting down a few notes, whistle or hum a tune, or say a few key words. Beyond these primary expressions, our draft statements are more elaborate, often involving writing, drawing, modeling, composing in a variety of media including paper, film, and the newly emerging electronic media. A trick to using our creativity is to learn to get ideas into a medium that we can work with. Previously, this trick involved mostly learning to get ideas down on paper. Mastering the tools of electronic media enables a freer flow. Creative thinking

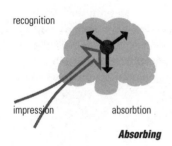

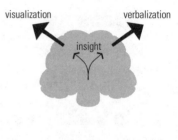

Absorbing   *Contemplating*   *Expressing*

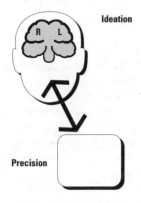

Couple mind and media

Channels for Expression Information

involves working with whole impressions. We can use visual images, symbols, feelings, and sounds, as well as verbal statements. Usually, our insights or hunches are an integration of all our life energy. Our first expression of an idea is often just a change in heart rate, or perhaps a smile (or a frown). Ideas flash through our mind at a tremendous rate. The quickest expressions of ideas are often verbal comments or visual notes. Producing primary expressions quickly reduces the chance of losing ideas. Writing down key words or making quick diagrams or idea sketches provides hooks for remembering insights.

Each of us has favorite ways of expressing ideas, certain transfers and media that we find comfortable. Yet with practice, we can all become comfortable using a wider range of media creatively. We also can learn to collaborate creatively with others expert in using different media.

Our imaginations are linked to all our senses. We derive images, emotions, and ideas from many aspects of our being. To find ways to transfer this energy, we need media that relate to many channels of expression. Each channel (such as writing, composing music, or drawing and painting) has art forms with long traditions. People specialize to master the knowledge required of a discipline and the complexity of whatever traditional media they use. Some—like Leonardo and Michelangelo—worked in multiple media centuries ago. For this they are known as Renaissance artists. Their work resonates with the universal force of life energy.

Because the tools available today are making it easier to work in multimedia, I believe a new Renaissance may be emerging, manifesting the wholeness of humanity. As more people become able to express their creative thinking, new art forms and products can emerge. Multimedia can make learning and collaboration more stimulating. Creative workgroups can coordinate their collective compositions, involving many modes of expression—sharing minds and knowledge—to build intellectual capital and a global consciousness.

23

Summary

Methods for Using New Media to Gather Information and Ideas

1. Set up an information environment where you can compile a digital design journal. It may be as simple as a word processing file in which you also can insert some graphics. Or it may be like a personal web site with a collection of written statements and graphics, as well as video and sound clips that you link together with a home page or some other simple catalog. You may want to eventually archive your journal on CD. Design journals have long been a part of creative endeavors. Now we have the opportunity to work interactively with multimedia in digital design journals.

2. Select your tools. Word processing works with text. Voice recognition transcribes words. Tape recorders capture sounds and conversations. Digital video cameras capture images as well as live action. Imaging programs enhance and edit photos. Scanners digitize printed material and quick sketches done by hand with pen or pencil. Illustration and graphics programs can be useful in developing visual notes. Select tools you can learn to comfortably use to record your perceptions

3. Use multimedia to take field notes. Work with familiar tools such as a pencil or pen on a note pad or sketchpad. Also take images using a digital camera or scan photos. Tape record sounds you hear and what people say. Use a video camera to take images you can annotate verbally, as well as full motion video with sound. Multimedia draws upon a wide range of perception and expression.

4. Collect multimedia information. Draw from books, electronic databases in libraries, CDs, and the World Wide Web. We are used to taking notes from books. It is just as easy to scan images, drawings, and maps. Identify sources for reference, as well as to acknowledge the author appropriately. Capturing written and graphic information from CDs and online is even easier. One way is to capture the screen image and digitally paste it into your journal. Another way is to transfer the file so you can work with the information.

5. Compile your field notes and other information you collect. You can organize your journal chronologically—like a diary or travel log—using dates. You can also compile your journal topically using key words or organizing themes. Digital design journals make it easy to rearrange and link information. Use this flexibility to advantage.

6. Add your interpretations and insights to your journal. Interpret the perceptions and information you are collecting, distilling the essence by adding your insights. Use both deductive and inductive modes of thought. Good journals both record information and stimulate the imagination. This is, after all, a source for innovation that helps you create better designs. Your goals, perceptions, and use of media will evolve.Seek meaning and value. Make sure you are satisfied with the quality before you release it online.

Activities

Use New Media for a Digital Design Journal

Design involves multi-sensory phenomena. The challenge is to gather information, capture perceptions, and develop ideas using multimedia. We can stimulate our imagination by working interactively with new media to compile a digital design journal.

Journals are a good way to keep a travel diary, recording observations and perceptions. (See John Muir's notebooks on Yosemite Valley done a hundred years ago.) Journals can be a way of compiling personal research, exploring ideas. (See Leonardo da Vinci sketch books done 500 years ago.) Today, digital tools make it easy to gather information in the field and online. Depending upon how you view the world, your journal might focus more on factual information and text, or it may focus more on perception and images. Work with both to draw upon different modes of thought.

Select a topic of personal interest for your digital design journal, or focus on the project you have selected for the first activity. Use the methods for gathering information, phenomena, and ideas already described.

If you are working as a team, clearly identify who is to focus on what— each person can contribute different perceptions that a team can use to advantage. Identify field work and research assignments that will be useful for the design project. Work out ways to assimilate information, ideas, and perceptions into your digital design journal using tools that are available to you. (A pencil and video camera are my field tools of choice.) Go online using the World Wide Web to gather information, images, and your own ideas that may emerge as you "surf the web." (The Design Professional Links Library identifies some useful sites.)

Consider other information sources—books from which you can glean notes, and scan images. There are CDs from which you can get digital base maps, images, and generic drawings for your endeavor.

Find clever ways for compiling, sifting, and linking the information to help generate fresh insights. Alternate between deductive and inductive thinking. Draw upon deeper levels of consciousness. Use reflective design practices to contemplate your perceptions and record your insights.

Much of your journal may be for your eyes only. Keep it as a record of the information, perceptions, and ideas you explored. But it can be useful to share information and insights. Share your journal with your alter egos—looking at it from different points of view. Share it with your collaborators—providing them with bits and pieces that would be useful to them online. Package key pieces for your client or an audience that could benefit from your work. Remember to select just what is most significant to others and make it accessible to them.

Use technology appropriately. A simple pencil sketch on a paper napkin may be ideal when discussing a design in a restaurant. Developing that sketch into a digital graphic that we can attach to e-mail, or place on a web site, is appropriate when collaborating online.

3

Make Digital Models to Represent Realities

Good models help us design more effectively. Models are representations of reality. They may be an actual prototype, virtual prototype, image, drawing, map, design criteria or, even a financial model done on a spreadsheet.

It is best to start off simply and then add complexity when building models. Just as a document may begin with a few key words, developing an outline, and then expanding into a draft, a spreadsheet model begins with a format, a few key formulas (which should be tested) and then information. A drawing or map should start with a few key organizing elements, which can be added to. Three-dimensional modeling, we can begin with a basic wire frame, a primitive of the object, or a base of the context that can serve as a frame of reference. These help set up the model quickly to see how well it will work before investing the time and effort to develop it. Although we want the model most appropriate for our purposes, in many situations simple models may suffice.

Frames of Reference for Computer Modeling

Three-dimensional modeling is complex, but it is also compelling, because it relates to the real world people perceive. In effect, design professionals can create virtual realities to develop and help implement designs. A simple three-dimensional model also can provide a frame of reference for linking more abstract models. The difficulty with three-dimensional models is that they take time and effort to develop; they also require more powerful computers to manipulate. Most design professionals still translate three-dimensional designs into orthographic drawings (plans, sections, and elevations) that can be plotted or printed on paper, creating scale-accurate drawings to build from. However, creative people are beginning to develop ways to use three-dimensional computer models to guide numeric-controlled tools—such as lathes or laser cutters—that fabricate physical components.

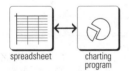

spreadsheet   charting
                     program

Link quantitative and graphical models

More abstract models sometimes have the advantage of representing phenomena we normally can't see. Systems models show relationships, while process models involve time. Quantitative models show calculations and patterns of statistical relationships. Many people use computers for quick and powerful abstract models. Combining verbal, quantitative and graphic models makes them easier to comprehend and present. We can do that using multimedia.

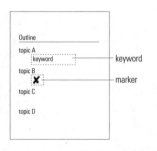

Word processing

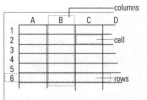

Spreadsheet

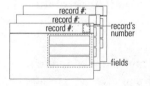

Database

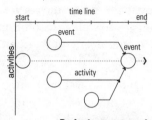

Project management

Design professionals need to understand frames of references to select and develop models. Here is an overview of generic computer programs with digital tools to build models for creative endeavors:

• *Word processing programs* are for doing narrative descriptions—such as a treatment, project proposal, program, performance criteria, or specification. Most word processing programs have outlining commands offering both detail and an overview. Using an outline in one part of a screen and the document in another can help us work with both the overview and details interactively. The "find" command quickly searches for key words.

• *Spreadsheet programs* are useful for doing cost estimates, energy budgets, water budgets as well as other calculations and matrixes. They organize information in rows and columns. This arrangement enables us to find cells of information very easily. We can embed formulas into this format to carry out calculations. A two-dimensional matrix provides a very handy frame of reference for a chart. It can handle information ranging from simple charts to complex ledgers or financial models. Working with spreadsheets, we can quantitatively model "if/then" situations to optimize decision making.

• *Database programs* enable us to sift and sort information interactively using search criteria. Some display information on a spreadsheet. Others use forms and a computer screen, much like stacks of file cards. Each file contains records and fields. The database can be searched by sifting and sorting these records and fields of information according to criteria we enter.

• *Project management* software graphically relates activities and events to a time line using a bar chart or work flow diagram to plan and manage projects. Lists assign activities indicating start and end dates. All this relates to a calendar. Project management programs also track resources and compile progress reports.

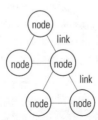

Hypermedia

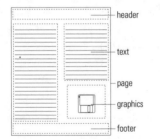

Desktop publishing

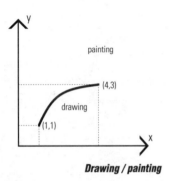

Drawing / painting

• *Hypermedia* provide "hot links" to associated information. Web pages and other hypermedia can be published online, linking multimedia for users to browse interactively. The author can map hypermedia as a network of choices. Each choice can nest more information.

• *Desktop publishing* software is for designing page layouts, integrating text and graphics. The page formats and numbers become the spatial reference, creating an electronic mockup, or prototype, of the printed document. The digital document can be laser printed or sent to a printing press.

• *Graphics programs* provide a two-dimensional frame of reference for visual compositions that may be representational drawings or abstract diagrams. There are drawing programs with symbols libraries and painting programs with color palettes. Illustration programs combine more graphic tools. Imaging programs include tools for working with photo images. Images can be digitized by using scanners, digital cameras, or digital video capture. Images provide their own frames of reference involving perspective and scale references as well as resolution, color, and contrast. The user navigates by zooming in and out, or panning around the graphic.

• *Computer-Aided Design and Drafting* (CADD) programs are for modeling form. Drafting programs provide a scale-accurate two-dimensional frame of reference with opportunities for layering information—overlaying different graphic information. Turning layers on and off can composite graphic information in different ways. Clear layering strategies help keep track of information and permit setting up base layers over which unique information can be drawn. That way several drawings can share base information from the same graphic file. Computer-aided design programs provide three-dimensional frames of reference as well as a coordinate systems for viewing the model. Attributes can be attached to objects, which provide links to databases. CADD systems also have parametric tools that provide another way to draw and explore forms, using key dimensions to generate drawings. Building three-dimensional models electronically creates a virtual prototype. When moving around a three-dimensional model, we need cues—like the ones we experience in reality—to orient and navigate. Design professionals can export models, or views of objects, into rendering programs to add colors and textures. Some artists now paint digitally in three dimensions.

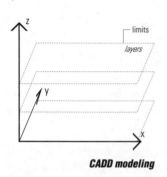

CADD modeling

• *Geographic information systems* (GIS) link spatial information, such as maps, with databases, such as land use, and develop spatial models using layers of information. Compositing information meeting certain criteria can generate suitability models. GIS links databases to spatial models, updating them as the spatial models change. It can be very important to relate spatial models to the earth's coordinates. That way the user can accurately relate the mapped information to the reality on the ground. Many of the digital graphic tools found in GIS systems are similar to those found in CADD systems.

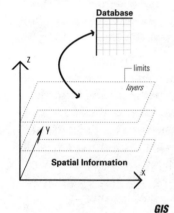

GIS

• *Animation* and *video* model movements through space in time. Most animation programs are two-dimensional, although some involve three-dimensional models. Storyboards can help visualize movement by planning positions and picturing the scenes. The development of in-betweens ("tweening") creates movement such as gestures. Movement is composed into flicks, sequences, and scenes in a time frame. An object can change by transforming from one morphology ("morphing") to another. These digital models are often used for special effects, to provide the dynamic qualities of movement and change. Video also represents full motion and can be modified using special effects software. Animation and video can be integrated by using green screens—providing a neutral background on which to superimpose images—and other techniques.

Animation

Music

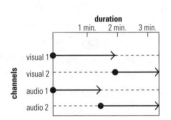

Multimedia

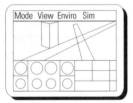

Flight simulator

Intrinsic Order

intrinsic order The natural or inherent order. Recognizing the intrinsic order of what you are working with enables you to model or represent it using a computer.

map To represent spatial order.

model To represent functional order.

• *Music programs* are used for audio production. This can involve musical notation for interactive composing, recording input digitally, sound sampling, synthesizing, editing, sequencing and mixing, as well as recording output onto master tapes or CDs. Soundtracks can also be integrated with video or film.

• *Multimedia programs* dynamically link segments of various media into a production. Working from a treatment, storyboards, and script, a creative director can integrate multimedia into time segments to tell a story. Digital tools help integrate different channels—including text, graphics, video, and sound—as well as create the transitions. The production may be broadcast or put on disk, video, or film.

• *Simulations* are another way of working with multimedia and are used for digital games and training. In effect, they can provide a virtual reality where the user has controls to interact with this environment. For example, flight simulators enable pilots to look out the window of a simulated airplane and view the horizon as well as the flight instruments. Simulators for training pilots include actual controls with tactile feedback, as well as seats that move to simulate kinesthetic sensations. Computer and video game versions of simulators use the keyboard, or various pointing devices—such as a joystick—for the controls.

Basing the models we use on the *intrinsic order* of the realities we work with draws more readily upon our intuition. For example, if we are working with an area of land, we can develop a digital map or three-dimensional model that represents that landscape. Viewing a spatial model provides us with a better sense of what is going on than would a chart numerically listing distances between points.

• *Mapping and spatial modeling* help sort out observations of phenomena that have a spatial or functional order. Using computers to work interactively with this information helps explore the intrinsic order. We can recognize textures, patterns, objects, spatial quality, orientation, sequences, and rhythms. This has interesting implications for the arts as well as for scientific inquiry.

30

• Database models enable us to search information that may not be locationally or functionally related. We can find patterns that are intrinsic to attributes we select. For example, a demographic database can chart the distribution of people in different age groups and relate it to a curve. Graphic representations of quantitative relationships and patterns can enhance awareness. For example, plotting costs and revenues shows the break-even point.

• Using evaluative computer models can check selected information and automatically carry out programmed responses. With programmed trading, computers evaluate stock market trends and then carry out transactions to buy or sell stock. Another programmed model is a quality control system that can measure attributes of parts and compare them to predetermined parameters. The computer can then quickly accept or reject these parts. In these models the programmer does the real thinking by predetermining what the responses should be.

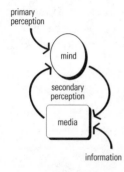

primary
perception

mind

secondary
perception

media

information

Primary and secondary perception

Digital models can enhance primary perception (our recognition) and aid secondary perception (our interpretation). However, we do the cognitive and conative thinking, as well as the judging, in our minds. Each of us determines what information means and what we are going to do about it. We recognize—and attach meanings to—images, written words, symbols, and patterns; we develop a sense of orientation that helps us discover and express new possibilities and make judgments to act upon them.

Multimedia offers richness because it relates to so many channels of perception—seeing, hearing, touching, and moving. Electronic media also provide many channels for expression—talking, writing, drawing, modeling, etc. As we learn to build multimedia computer models, we can, in effect, learn to create virtual reality.

Design professionals work with many frames of reference. As the content providers, we can shape different types of models using information environments. Operating systems and communication programs help integrate a vast array of digital tools that enable us to do that. Models enable us to work more interactively with information so that we can refine and test ideas before they are carried out. We can also recycle information and the products of creative energy. Since a model may mockup what we are producing, it also enables us to develop production procedures concurrently with the design. These models become an artificial reality that a project team can work on together in a shared information environment.

Abstractions

Abstractions simplify information and clarify relationships we perceive, enabling us to understand larger patterns not readily comprehended at concrete levels. Abstractions also help express and develop concepts. We can work with abstractions using any medium to relate different channels of perception and expression.

Written language has different levels of abstraction. For example, in reality, you may see and smell a cow. Call her *Bessie*. The word *cow* abstracts general characteristics commonly related to *Bessie*. Referring to cows as *livestock* further abstracts their characteristics. Livestock relates to farm assets, which in a more abstract sense are just assets. At an even more abstract level, is the term wealth, which *Bessie* represents. Each level of abstraction leaves out certain information, but enables us to consider further understandings.

In his book *Experiences in Visual Thinking*, Robert McKim shows how to use graphic languages both concretely and abstractly for visual thinking and graphic expression. He identifies concrete graphic languages as involving working models, three-dimensional mockups, perspective projections, isometric and oblique projections, and orthographic projections. Abstract graphic languages include schematic diagrams, graphs, and charts.

abstraction *An operation of the mind involving the act of separating parts or properties of complex objects. Enables you to simplify information and clarify relationships you perceive.*

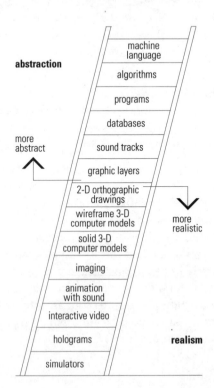

Multimedia abstraction ladder

New media also relate to different levels of abstraction. Design professionals can move ideas up and down a multimedia abstraction ladder. The most concrete level may be physical prototypes or models cut with numerically controlled tools. At the next level is virtual reality experienced in simulators that provide visual, audio, tactile, and kinesthetic feedback. At the next level there are interactive video and audio and then animation with sound. Three-dimensional models are more abstract; they range from rendered solid models to wire-frame models. Beyond that there are two-dimensional orthographic drawings such as the plans, sections, and elevations typical of most architectural and engineering construction documents. At more abstract levels yet are diagrams on graphic overlays. Soundtrack channels are like abstracted layers. Still more abstract are spreadsheet models and attribute databases. Beyond that are programs using object-oriented languages. Probably the most abstract level of electronic media are the machine languages that computers use. Programmers usually have compilers to attain that level of abstraction. Few people can program in machine language.

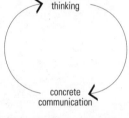

Alternating between abstract and concrete

When designing, we can use new media in both abstract and concrete ways. Abstractions help interpret and express thoughts. They provide quick ways of sharing insights to use as starting points when developing ideas. We may find ourselves most comfortable sketching and writing notes on paper and, initially, not composing at the computer. It is often more convenient to jot down observations or insights on paper whenever they occur. Once we have a digital model, and the tools to work with it interactively, it can become quicker to express and develop ideas in the computer. The concrete representations that new media provide can help communications.

There are advantages to abstract thinking. It helps simplify information and focus on key relationships. Abstractions help us quickly explore alternatives. Using multimedia to present abstract ideas can help comprehension.

There are also advantages to concrete thinking. Concrete thinking helps us examine details and communicate what designs will be like. Virtual reality works best for developing thoughts more concretely and for design communication. A realistic presentation can provide enough detail for other people to experience the design so they can judge it themselves by abstracting their own understandings and responses.

Cognitive Maps

Cognitive maps can clarify and express concepts and interrelationships. By expressing understandings visually, we can see relationships that may not be evident when just working symbolically using words or numbers. Cognitive maps can make it easier to relate information to

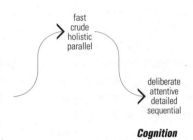

Cognition

the project at hand. This can help us visualize connections, often providing creative sparks. Concept diagrams help express new insights. Being able to communicate understandings and insights enables design professionals to collaborate more effectively.

There are many approaches to doing cognitive maps and concept diagrams. Ulric Neisser points out in his book, *Cognitive Psychology*, two stages related to cognition. The first is "fast, crude, holistic, and parallel." The second is "deliberate, attentive, detailed, and sequential." Neisser observes that cognition involves both abstract and concrete observations. We can go from abstract notions to working out details. We can also derive abstractions from concrete observations.

bubble diagram *A graphic representation that shows the relationships between functional areas. It can describe patterns evident in a map or a plan.*

flow diagram *A graphic representation that shows the sequence of a process. It can describe movement of material or energy in natural processes.*

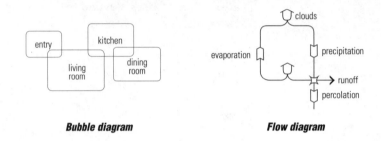

Bubble diagram   **Flow diagram**

Bubble diagrams can show the relationships among functions—such as land uses or areas of a building. *Flow diagrams* can show movements of energy or material—such as nutrient flow in natural processes or revenue flow in human processes. Flow diagrams can model food chains or decision-making processes. *Form and structure diagrams* can show the anatomy of what we are working with to understand archetypal forms—such as concentric or axial arrangements—as well as structural concepts—such as a cantilever, span, or arch. We can develop concept diagrams that reflect the intrinsic order of what we are working on. Initially our diagrams may be fast and crude. As we develop our concepts, these models can become more deliberate and detailed.

Today, digital tools are emerging which enable design professionals to use graphic languages to manipulate many patterns and processes. Graphic user interfaces—with icons as well as graphic diagrams for navigating—are part of most computer applications. Project management software generates work flow diagrams. Multimedia authoring programs provide icons on a timeline to integrate different channels into presentations. Flow diagramming programs provide symbols libraries for diagramming many types of systems.

Hierarchical Orders

Examining *hierarchies* also helps us understand the intrinsic order of what we are working with. Hierarchies can be categorical, spatial and/or temporal, and systemic. Most hierarchies exist in a continuum without boundaries. That is, there are entities within entities. We can work with hierarchies by establishing frames of reference when building our models.

Outlines—often based on topical categories with hierarchies—provide both an overview and detail. Using hypermedia we also create links that have hierarchical orders. Web navigation tools enable us to navigate links of information thus going deeper into detail.

Spatial models also have hierarchies. When using spatial models, we need to recognize the scale of concern that we are dealing with, then consider at least one scale above and one scale below our level of concern. For example, when modeling a site, map the vicinity as well as specific locations within the site. We need to make sure that the accuracy of the map is appropriate for the scale.

Temporal models have hierarchical time frames. We should recognize the time frame we are working with. For example, when doing short-term planning, we should also consider the long term as well as the immediate future to provide continuity. There are also temporal hierarchies in audio and video productions.

Systems have hierarchies too. Transformations result in processes that make up systems. We can show transformations as black boxes. Linking the inputs and outputs of transformations, we can describe processes. Linking processes, we describe systems. Here again, we need to recognize the level of understanding that is most useful for our purposes. For example, when designing an irrigation system we should show the point of connection, backflow device, and valve circuits. Also, we should look at the next higher level of function. For example, where is the water coming from? Where necessary, look at lower levels to understand how the parts of the system work. Computer graphic or flow diagramming software can help us do that.

Hierarchical orders are built into many computer applications. Recognizing them enables us to work with these applications more easily and more creatively.

Parametrics

Parameters are key variables that govern the shape or performance of models. CADD programs use parameters to generate forms. For example, simple offsets can define setbacks or road widths. Arrays can define column spacing. The raise and run of a step can generate the form of a stairway. Engineering design programs also make extensive use of parameters. Parameters, such as proximity, are also used for spatial modeling in GIS. For example, a GIS program can display all the locations within a specified distance from a highway or stream. The challenge is to find key parameters that can be form generators and to use them creatively.

Smart Drawings

The polygons that make up computer graphics have properties such as areas that are useful for cost takeoffs and other queries. We also can link objects to *attributes* by attaching database information. In effect, we can create smart drawings with considerable embodied information. (A drawing of a layout of objects could link to a database with the manufacturer, model, and cost of each object.) Geographic information systems link to more extensive databases. Many communities are compiling geographic data on public property and utility infrastructure. These data can include both locational information for mapping and database attributes for tracking key information. Linking maps and attributes enables computer users to keep track of resources in ways that were not possible when using only hand-drawn maps.

Summarizing Techniques

Digital models provide frames of reference that enable us to use visualization techniques, enhancing our perception for design. Multimedia draws upon different modes of thought. Abstractions help sort out meaning and clarify relationships. Cognitive mapping transfers insights and understandings into visual forms we can work with. Hierarchies address scope and scale. Parameters help develop designs according to set criteria and explore options. Smart drawings link to databases integrating nonvisual information.

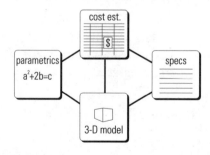

Link models

Realities

Model space is a way of referring to the information environment where we work on our models. Another reality we deal with is the presentation reality—what others will perceive. This may involve the pages or drawings we produce on paper (paper space) or animated images and sound we produce on video or online (screen space). Artificial reality and presentation reality may be one and the same if our audience can experience and directly interact with what we are working on. More typically, we transfer information from our model to some presentation format—pages on the web, printouts on paper, slides, film, or video sequences on tape or compact disk. We can select views of the artificial reality we are working with and package them for an audience.

model space *The information environment where you work on your models.*

presentation reality *The information environment others will perceive. This may involve the layout of pages or drawings you produce on paper, or it might be the sequence of a video production.*

paper space *A representation showing what you will see when you print a document on paper.*

screen space *What you see on a typical computer monitor. The display may vary unless laid out in a standard screen format such as Adobe Acrobat.*

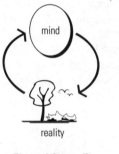

Recognizing reality

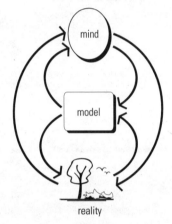

Representing reality

As design professionals we learn to work with the dichotomy between the real world and the world we can imagine. With computers the distinction between reality and artificial realities is becoming blurred. Reality in physical space can relate to virtual reality in cyberspace. It is becoming possible to trick the viewer into thinking an image is real. That poses ethical questions. Our use of media should not be deceiving. It may be important to maintain some unreal qualities as a characteristic of the media we use.

Summary
Methods for Modeling Realities

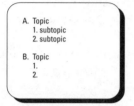

Outline

Format

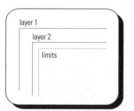

Graphic layers and limits

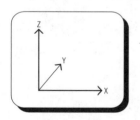

Three-dimensional modeling

1. **Identify key realities you are dealing with and determine what models would be useful.** Draw upon the digital design journal you compiled in the previous activity. Consider models that would help compile and analyze information, as well as synthesize and communicate design. These may be diagrams, maps, drawings, and design criteria typical of traditional media. The models may also include spreadsheets, hypermedia, imaging, CADD and GIS, as well virtual reality using new media.

2. **Determine which digital tools would be the best for building these models.** Select the hardware and software you have access to that would be most helpful for your project.

3. **Build analytical and interpretive models.** Begin with information, perceptions, and insights gathered in your digital design journal. Develop models to absorb, contemplate, and express ideas. Use abstractions to clarify your understanding of the problems and concepts for the design.

4. **Share models of key realities with consultants, with your client and constituency (or market), as well as with review bodies.** Draw upon the expertise necessary to help develop useful models to find out more about the problems and opportunities. Share the models with others to get their perceptions. Collaborate through meetings, phone calls, and online interaction using models of the realities you are working with.

5. **Verify and refine the models to reflect or interpret the current situation by spending time with the site or situation.** Develop a focus. Models can simply be a means for generating a design. Digital models, however, may have a life of their own in situations where they can continue to be useful, or have particular interest or value to people—such as with evolving public work.

6. **Derive insights by drawing upon a repertoire of thinking skills.** Share these insights with others to see what they think and how they feel about the problems and opportunities. Remember, you can do much of this online, tapping into a collective consciousness. Let your design journal and digital models evolve interactively. They will become intertwined.

7. **Develop goals and objectives as well as design criteria.** Design guidelines based on your modeling and interpretation of the realities you are dealing with will help you move from analysis to synthesis.

Activities

Make Digital Models to Represent Key Realities of a Project

The challenge is to build models of the key realities you work with to help analyze problems and synthesize new possibilities.

Select a project for which you would like to develop digital models. Work collaboratively with your design team to develop and refine your models using the methods described in this chapter.

Use electronic spreadsheets if you need quantitative models addressing costs and revenues as well as energy and water budgets. Systems diagrams can model information flow, workflow, and revenue flow. Use systems models to examine material and energy flows in natural processes for regenerative designs that close material cycles and use renewable energy. Surveys can provide digital terrain models of landforms to use in CADD as well as GIS, which can link spatial attributes. If your project involves imaging, the key realities to model would be the colors and textures defining the visual character of what you are working with. You could combine the image with a three-dimensional model to study the form of an object or spatial qualities of an environment. Animation, or three-dimensional models, help examine how the object moves, or how one moves through the design. You might even produce a video showing what this would look like.

Maybe you are already doing digital plans, sections, and elevations of your design using CADD. Can you go further, using three-dimensional models of the context and design? Can you link attributes to the components of the design and model costs, as well as energy or water budgets using databases and spreadsheets?

Draw upon references and consultants to develop the necessary models. Archives of digital information are growing online, and on CD—providing everything from maps and images of large land coverages to drawings of small details that can help you build digital models.

Set up the models so that you, and your design team, can use them online to interactively analyze the realities you are working with. You can send them attached to e-mail or place them at a web site. Use these models to help explain the site or situation to your client or constituency. Your analytical models are the basis for design.

Linking data to objects and locations

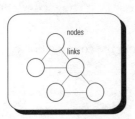

Hypermedia

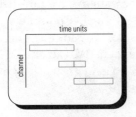

Scripting environment

4

Design with Digital Tools

Visualization is a key to creativity. We can use digital models to express, test, and optimize designs, but to use these effectively, we need to link what we experience in artificial reality with our experiences in reality. When looking at digital drawings or design models, we should be able to see not only what is on the screen, but imagine what is at the site or situation. Immersing ourselves in reality—as well as in artificial reality—enables us to draw more deeply upon our senses and intuition.

Pattern Seeking

The capacity to recognize patterns helps us comprehend what we perceive. Patterns enable us to visualize new possibilities and perform activities such as using digital design models. Here is a method for doing this:

1. *Survey* the whole image or model to gain an overview by perceiving the patterns.

2. *Focus* on questions to pique your interest.

3. *Pace yourself* letting your attention follow the flow of your thoughts. Avoid getting bogged down or distracted.

4. *Speculate,* visualizing connections to your past knowledge and experience.

5. *Instantly access information* by visualizing the composition of the image or the organization of the model.

6. *Sharpen* your analysis by visualizing patterns to organize new information and see if it fits this intrinsic order.

Linear and Lateral Thinking

Typically, we work most comfortably with *dichotomies*—sets of two (usually opposites)—or *trilogies*—sets of three. With practice we probably can work (in our mind) with about nine patterns at a time before we get confused. (Try visualizing three sets of three.) Computers enable us to work with more possibilities, since we can store many patterns, recall them easily, and manipulate them. We can add or update information, as well as restructure it to gain greater insight for visualizing new possibilities.

Dichotomy   **Trilogy**

dichotomy *A division into two parts. A set of two (usually opposites).*

trilogy *A discourse consisting of three parts. A set of three.*

po *A positive maybe.*

vertical thinking *An approach to creative thinking typically involving a linear, logical progression of steps.*

lateral thinking *An approach to creative thinking typically involving consideration of alternatives.*

In his book, *Lateral Thinking: Creativity Step by Step*, Edward de Bono uses the word po (a positive maybe) as a way of identifying possibilities. Linear thinking usually deals with *yes* or *no*. Lateral thinking adds *po*—a positive maybe—to keep possibilities alive.

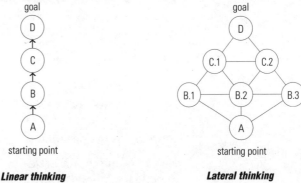

Linear thinking   **Lateral thinking**

Of course, computers are marvelous machines for vertical thinking. They process data using programs based on logical modes of thought: *on-off, yes-no, if-then*. By working interactively with new media, however, we can explore new possibilities—artificial reality is the *land of po*. Some of the most exciting possibilities for using digital tools creatively have to do with applying lateral thinking. Working interactively, we can restructure patterns, recognize new possibilities, and quickly manipulate them. In this way, computers can enhance both vertical and lateral thinking.

Thinking Analogies

We often develop our comprehension of similarities in the form of analogies. This capacity helps explore relationships and make creative connections. In the book *Synectics: The Development of Creative Capacity*, William J. J. Gordon identifies four types of analogies useful for creative thinking. These are symbolic analogies, direct analogies, personal analogies, and fantasy analogies.

Symbolic analogies involve abstract qualities that relate from one situation to another. For example, the *form* of the Sydney Opera House, in Sydney, Australia (designed by Jorn Utson), abstractly appears like sails in the Sydney Harbor. Symbolic analogies can also involve abstract concepts. (The Statue of Liberty commemorates the freedoms embodied by both the French and the American revolutions.) Symbols are very powerful analogies useful not only for design and art, but also for political action.

Sydney Opera House

Statue of Liberty

Bird in flight

Direct analogies involve similar physical characteristics or processes that relate to different contexts. We can use direct analogies to transfer concepts from one situation to another.

Nature is one common source of direct analogy. (A bird's wing is directly analogous to an airplane wing.) Centuries ago, Leonardo da Vinci filled sketchbooks with observations of nature. By exploring natural phenomena, he derived direct analogies that provided the basis for many of his inventions.

Raindrops

Personal analogies relate to our identification with elements of a problem. By pretending to be a raindrop, we can look at rain clouds in a more penetrating way. Personal analogies also enable us to project ourselves into another person's point of view and develop empathy for that person's situation.

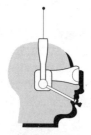

Fantasy headset

Finally, Gordon identifies the *fantasy analogy*, which describes an ideal. Although this analogy is detached from reality, it can provide a source for visualizing real possibilities. For example, let us start with the outrageous fantasy of implanting computer chips in our brains to enhance our mental capabilities. Using today's technology, this fantasy might be developed into a headset—something like a jogger's radio or a helmet. In addition to having stereo and TV, this fantasy headset also might include a miniature computer and a portable telephone. Voice

42

might be the main mode of interaction. Eye movements could operate pointing devices, keeping hands free to do other tasks. A headband or helmet might even detect temperature and galvanic skin response to know how the user is doing and help induce a good frame of mind. Earphones with surround-sound coupled with three-dimensional visual-display eyeglasses could immerse the user in virtual reality.

Fantasy transcends the boundaries of reality, since, in effect, the only boundaries are the limits of our imagination.

> *"Imagination is more important than knowledge."*
> — Albert Einstein

Transformations

transformation *A change, or visualization of change.*

visualization *The formation a mental image that can help you coordinate your mental and physical activities.*

Transformations help visualize change by putting models together (construction) or taking them apart (deconstruction) to explore new possibilities. We can make transformations concretely using physical forms or abstractly using concepts.

Digital models can be put together or taken apart rather easily (certainly much more easily than a scale model or an actual prototype). This can reduce the time it takes to research and develop designs.

Interaction

Not only does digital visualization help coordinate mental and physical activities, it also helps us draw upon our creative energy and direct it toward addressing the problems at hand, coordinating mind and movement.

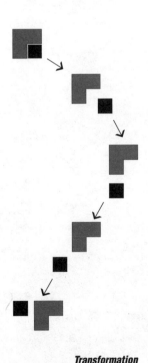

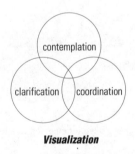

Visualization

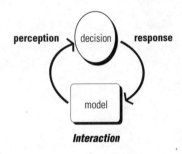

Interaction

Transformation

The basic approach for interacting with computer models is similar to the interaction we enjoy in games or sporting activities. We perceive situations through our senses, and these perceptions provide a basis for our decisions on how to respond. Each response brings about a new situation to which we can respond again. Eastern disciplines like aikido bring interactions related to self-defense to an art form.

Interaction can absorb our whole being, engaging us again and again with each new situation that presents itself on our computer screen. We can relate reality to artificial reality. Visualizing what we are going to do and how we are going to do it helps focus creative energies. We can learn a repertoire of techniques or moves that can be exhilarating. Each type of software embodies digital tools that we can make part of our repertoire. With practice we can use them interactively for creative endeavors.

Collaboration involves a design team working together in a coordinated fashion—as in dancing, people can work in harmony, anticipating each other's moves. Online communication provides new opportunities for partnering. Online, each design professional can creatively contribute to a collaborative effort involving multi-individual expertise. To collaborate effectively, we need to understand the creative process.

Creative Process

creative process *The stages your mind goes through when developing ideas. These stages include preparation (involving both first insight and saturation), incubation, illumination, and verification.*

We can nurture our own creativity, and the creative contributions of others, by understanding the thinking process we go through when developing ideas. This process, which involves problem stating and problem solving, includes five stages:

1. First insight

2. Saturation

3. Incubation

4. Illumination

5. Verification

Design Process	Comparative Description of the Creative Process								Learning Process
Various Disciplines	Helmholtz	Poincaré	Getzel	Edwards	Lowenfeld	Gordon	Halprin	McKim	Whitehead
			first insight	first insight					romantic stage
research	saturation	saturation	saturation	saturation	preparation	problem stating	resources		
analysis	incubation	incubation	incubation	incubation	incubation		scoring		precision
synthesis	illumination	illumination	the Ah-Ha	illumination	illumination	problem solving	valuation	express	
evaluation		verification	verification	verification	verification		performance	test cycle	generalization

Each stage may take varying lengths of time, depending upon what we are doing. Betty Edwards—artist, educator, and writer—provides descriptions of the process in her book *Drawing on the Artist Within*. These descriptions and others are summarized in the chart. From this chart we can see the similarities. The terms tend to overlap. Working through the transitions between the stages is crucial. This enables us to progress through the creative process.

RSVP cycles

The creative process is *cyclical* rather than linear. The landscape architect Lawrence Halprin pointed this out in his book *RSVP Cycles: Creative Processes in the Human Environment*. Halprin developed his own terms for describing that cyclical process, but these correspond to terms that have been used by others. The *R* in *RSVP* stands for *resources* (inventorying information, developing goals, motivating ourselves)—which is a form of preparation. The *S* stands for *scoring*, which involves diagramming and symbolizing what we are working with. This is more than just incubation. Halprin considers scores to be like musical scores. The *V* stands for *valuaction*—a word Halprin coined by combining the words *value* and *action*, key notions involved in illumination. Finally, the *P* stands for *performance*—involving the implementation or testing of the idea, which is essentially verification.

People typically go through a cyclical creative process in iterations. Robert McKim, in *Experiences in Visual Thinking*, writes of *expressing, testing,* and *cycling* ideas, which he refers to as *ETC*. The cycle involves first expressing and then testing ideas. One idea can lead to another. We can go through this cycle frequently, each time adding refinement to our ideas, with each re-iteration.

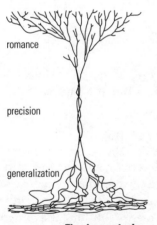

The river metaphor

Learning is related to the creative process. Alfred North Whitehead described the learning process as involving three stages: *romance, precision, and generalization*. The first insight is a *romantic stage* of the process. Saturation and incubation involve more *precision* as we examine information that is relevant. Illumination and verification can become a more *general* stage of the process where we make discoveries and determine how to apply ideas.

In his book *Design for Human Ecosystems*, John Lyle related Whitehead's description to the design process by using a river as a metaphor. He pointed out that the origin of a design flows from many sources during the romantic stage. The flow becomes more directed during the precision stage. Finally, like a river flowing into a delta, designs branch out during the generalization stage, involving synthesis of new possibilities.

The Design Process

The design process used in disciplines such as architecture and landscape architecture usually involves:

1. Research (problem identification and information gathering)

2. Analysis

3. Synthesis

4. Evaluation

design process *The stages your mind goes through when developing design ideas. These stages -- which are related to the stages of the creative process -- include research (involving both problem identification and information gathering), analysis, synthesis, and evaluation.*

This design process is really patterned after the creative process. When collaborating, we should recognize that each discipline typically has its own rubric for describing the creative process. The approach may vary somewhat and the words used to describe the approach may be somewhat different, but the basic pattern is similar.

Creativity in Each Phase of the Project

Design professionals, such as architects, go through phases when doing projects. The American Institute of Architects identifies the phases of the basic services that architects provide as:

Predesign

Preliminary design

Design development

Construction documents

Bidding

Construction administration

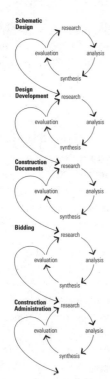

Intertwining creativity and design

The stages of the *creative process* and the phases of design are intertwined, for each phase of a design project involves the creative process.

During the predesign phase of a project, design professionals go through research, analysis, synthesis, and evaluation—defining the problems and opportunities, identifying objectives, and beginning to develop the program for the design. During the preliminary or schematic design phase, design professionals will again go through research, analysis, synthesis, and evaluation—establishing the design concept and basic forms. Then, during the design development phase, design professionals will once again go through the creative process— refining the design and adding more information (such as selecting materials). During the construction document phase, design professionals work out the precise layout and details showing how to build the project. Bidding involves still other iterations of the creative process, this time finding appropriate bidders and communicating the project clearly for them to be able to research, analyze, and come up with construction costs which are evaluated by owners and other investors. Even construction administration will include problems that need to be solved. Research, analysis, synthesis of alternative solutions, and their presentation might take place between job meetings or with great urgency on the spot.

The creative process involves every phase of a project's design. The focus in the initial phases is on design; in the later phases, on production. Each phase involves creative thinking and design communication—which we can do using digital tools with new media.

Creativity is also involved in management. Good management involves developing insights about problems in an organization or facility. People need to immerse themselves in these problems—saturation helps understanding. Incubation involves reflecting on what is going on. Illumination provides ideas to improve the management of an organization or facility. Verification is needed to see if the management idea will work. Clear communication will effect change. New media and digital tools can help organizations and businesses manage their endeavors.

Here are some guidelines to assist design professionals and managers in working with new media more creatively:

• Use new media to work collaboratively and also engage clients and constituency in design communication.

• Develop useful models to visualize the realities of the situation as well as design alternatives.

• Focus on doing what is necessary. Try not to get carried away with doing something just because there are tools do it.

• Use information appropriately—not just because it is available.

• Use media effectively throughout the creative process by selecting a combination of hand and digital tools that quickly transfers ideas and information.

• Work interactively, using the creative process to guide collaborative efforts.

• Relate to both multimedia models and reality to stimulate insights.

• Move freely beyond analysis into synthesis.

• Generate more than one idea quickly using conceptual models.

• Avoid prejudgment. Keep your mind open to possibilities by postponing judgment until there is a need to verify ideas.

• Focus creative energy on developing the most promising ideas.

• Use models to test designs before producing them.

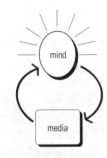

Interaction of mind and media

"In every work of genius we recognize our own rejected thought; they come back to us with a certain alienated majesty."
—Ralph Waldo Emerson,
*Self Reliance,*1844

• Derive further insights from testing models that will help refine the design.

• Connect with the audience by using new media for design communication that has both excellent content and positive emotional impact.

• Conserve creative energy by using digital tools to do more with less effort.

Ways to Visualize the Creative Process

A key to working out our creative process is to visualize our approach. This helps conceive alternative strategies, involve new tools and media, evaluate how to allocate time and resources, and manage our project as we proceed.

Probably the simplest way to visualize our approach is to make a *checklist* of activities. Many people call this a "to do list." We can simply write this list on the back of an envelope—but by using e-mail we can share this list with clients and the project team, quickly determining who will be doing what. A digital list is easy to add to and update as we interact with others online. The list of activities may become the scope of work in a contract, as we then identify what will be done at each stage of the project. We can use this list to record activities we have accomplished, as well as the activities we plan for the future. This list also becomes useful for developing similar projects. It can also identify "due dates" that help keep the project on schedule.

A *Gannt chart* can relate a list of activities to a project's time line. Activities, listed on the left side of the chart can relate to bar graphs showing the beginning and end of each task. In this way we can graphically see the duration of tasks in relationship to each other. This is useful for planning projects and managing progress. However, a Gannt chart does not really show the interconnections of the activities.

Gannt chart *A bar graph that shows the duration of tasks and relates them to a time line.*

work flow diagram *A diagram linking activities and events and relating them to a time line. Used widely for operations research and project management.*

Checklist:
> top priority
> second priority
> third priority

Journal:
> record
> ――――――
> ――――――
> agenda
> ――――――
> ――――――
> plan
> ――――――
> ――――――

Lists

Calendar

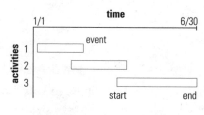

Gannt chart

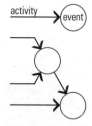

Work flow diagram

1. List activities.

2. Link activities.

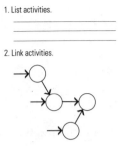

3. Refine procedures.

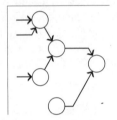

4. Develop work flow diagram.

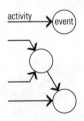

5. Revise & re-define. If necessary, change scope or change time frame.

6. Manage project. Update workflow diagram as you proceed.

Procedure

activity ⟶ (event)

Manage project

A workflow diagram can help plan and coordinate efforts by visualizing how activities and events relate to each other through the design process and phases of a project. To make a workflow diagram, first identify the activities and events involved. Then link them in ways that show what precedes and follows each activity. Consider how the procedures relate to each stage of the design process and each phase of the project.

The *critical path* is the sequence of activities that will take the longest time to accomplish. Consider ways to shorten the critical path. Diagram alternative approaches that incorporate different strategies for using available resources. Relate these procedures to a time line. Use the diagram as a focus for coordinating the team's efforts, identifying the paths for each team member to pursue. Clearly identify stopping points and milestones necessary to coordinate the hand-offs of information.

Use workflow diagrams to help manage projects. Update diagrams, recording progress and reflecting better strategies for proceeding. A workflow diagram can help determine what the schedule and budget should be to successfully complete a project. It can also help determine ways to complete a project within a budget parameter and deadline.

There are two major ways to use the design process diagrams:

1. As a general guide—keeping options open, or

2. With more precision—endeavoring to pin down key events.

Different situations will dictate which approach is most appropriate. In addition to helping us sort out our own approach, a workflow diagram should serve colleagues and clients. Keep it simple. There are limits to how precisely we can plan creative projects. Diagramming can help us think through ways to design with digital tools, but inevitably there will be much more to discover as we proceed.

Remember, we can diagram design processes either by hand (using pencil and paper), with computer graphics (using a drawing program), or with project management software. Project management software can help establish a calendar, develop workflow diagrams, and track resources with more precision. This is especially useful for large projects with elaborate teams. When calculating compensation, consider not only the personnel involved, but also the computers and other digital tools needed to work with electronic media.

Effective management involves taking the time to review progress. It helps collaborative workgroups assign responsibilities and coordinate efforts to keep design projects on track.

Summary

Methods for Designing with Digital Tools

1. **Refine the design program.** Use guidelines to establish key parameters (such as sizes and budgets) and other important design criteria, working from analytical models and your design journal.

2. **Clarify your design process.** List tasks online identifying who is doing what and when. Use a workflow diagram to manage the project, helping each member of the team see where he or she fits into the collaborative effort.

3. **Explore design concepts by building conceptual models.** Seek new patterns to generate alternatives. Use lateral thinking and analogies. Transform models. The models could be bubble diagrams showing spatial relationships, flow diagrams showing processes, or other shorthand concept notations. Sketches express ideas quickly. Digital concept diagrams help transform and communicate ideas. Use the strengths of each of these modes of expression.

4. **Develop design models based on the strongest concepts.** The digital design model can use base information from the analytical models. Develop both abstract and concrete models. Each team member may be responsible for a model. Interactively use design models as a focus for collaboration by having the design team work together, marking up and refining the models. Different modes of interaction—meetings, phone/fax, e-mail, and web sites—can help the design team collaborate and coordinate its efforts.

5. **Transform digital design models.** Save versions of the model at each phase of design. Transform the concept model into a preliminary design. Develop the preliminary design adding detail, which eventually becomes part of the implementation documents. All disciplines can apply this process using their digital design models.

6. **Test and optimize the digital design models.** Use them in the review process by printing them out for marking up, or by engaging the client and review bodies in working interactively with the digital design models online. Using e-mail and establishing a web site can help make the digital design models more accessible for review. Clients typically respond with comments, plan checkers with corrections. Use design models that provide a sense of what the design will be like to survey users and sample the market. Relate the digital design model to the actual conditions. Use all this feedback to refine the design.

Activities

Transform Designs Digitally

The challenge is to work with digital design models to reduce development time while improving design quality. Make digital design models a focus for collaboration by working interactively using new media. Transform design concepts into digital design models to share for evaluation and testing. Good models enable design professionals to creatively use digital tools to express, test, optimize, and even build designs. By working interactively and collaboratively with these models, a design can go through many iterations of refinement and benefit from the expertise of many people.

Build a team to develop design models for your project using the methods presented in this chapter. Use information technology you have available to engage your design team in a collaborative effort. Conceptual models help explore alternatives and explain concepts, providing a focus that is helpful in selecting design alternatives.

Use these models as a focus for collaboration. Make a printout for others to mark up with colored pencil. Invite team members to sit around a computer and interact with the digital diagrams, drawings, images, and forms to explore new possibilities together. Distribute base information digitally, enabling others on your team to add their contributions to the digital design model by working on different layers or components. Collaborate online so you don't always have to find a common time and place to meet.

Begin modeling designs, deriving insights from digital design journals and information from the analytical models of key realities you are working with. Analytical models can become the basis for building schematic design models.

Flesh out design concepts by adding more precision and detail. You may quickly sketch ideas by hand—using paper and pencil—and then transfer the sketchy ideas to digital models to test and optimize the design. As you become more comfortable with digital tools, it may become quicker to express, test, and optimize design ideas by working interactively with the digital design model. You may also work with others who can help transform your sketches into digital models, or who have the capacity to build useful digital models your team can work with.

Design models will evolve, building upon the base information and layers of ideas. There are many types of models your team may share. Have people on the team be responsible for building a model related to their expertise. For example, working with imaging can transform the colors and textures into the qualities that define the visual character of the design. Systems models examine connections and how the design works. Modeling the flows of material and energy helps to visualize how to close the cycles and link them with natural processes for regenerative design. Three-dimensional modeling helps transform the forms and spaces that express the functions and provide desirable spatial qualities. GIS models can help optimize locational decisions. Use digital models to design—interactively exploring alternatives and building up the layers of information. Save versions of the digital models as a record and back up at each phase of design. The team may want to revisit some early ideas. Remember that digital models are useful for design communication with clients, review bodies, and the prospective users or market for the design.

51

5

Communicate with New Media

Digital tools enable design professionals to share base information, collaborate with consultants, and communicate designs to clients and others using new media. Information technology also provides opportunities to interact with constituencies and customers—the users of the design in the marketplace—connecting with them in ways that can add value.

Computer technology is moving beyond automating routines. It is transforming into robust information technology that can augment and enhance communication. Using this information technology, design professionals can apply a shared knowledge base. Even more important in times of rapid change is the potential to improve the capacity for teaching and learning. This is especially true for creative endeavors that constantly chart new intellectual territory. Working online, we can learn by collaboratively expressing, testing, and refining ideas—interacting with clients and other constituencies who can participate in design. Information technology provides the tools to quickly implement designs and improve their quality. Opportunities to use new media for design communication have exciting potential!

Audience

Design professionals carry out many roles, and for each role there is an audience. Design consultants have clients and collaborators, design staff have organizations to relate to, researchers have funding agencies, advocates have constituents, artists have patrons, entrepreneurs have investors and markets. It is important to know who our audiences are—and how we can reach them—for effective design communication. Often we must package presentations for different groups. New media can help us do that more effectively.

My professional practice, the Claremont Environmental Design Group, Inc. (CEDG), is involved in a project for an orphanage in Vietnam. Certainly, the orphans and staff at this facility are an important audience, but our design communication also must reach many other people. We are part of a collaborative team—with the Geographic Planning Collaborative (GPC)—involving design professionals and technical consultants, working through a benefactor from Japan. We are proposing regenerative designs for producing electricity, food, and water from local renewable resources so that the orphanage can expand and enhance the quality of life. The initial focus of our design communication is to provide a vision for "Living Energy Systems" to help local

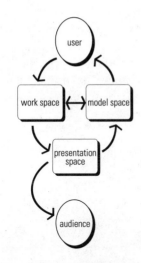

Merge work space, models, and presentation

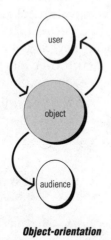

Object-orientation

governmental officials in Vietnam apply for international funding. In addition to obtaining funding from economic development and humanitarian organizations, we hope to communicate with environmental organizations interested in reducing global greenhouse gases to share this project as a model for other countries. Various governmental and business organizations are promoting the emerging technologies—photovoltaics, electrolyzers, reformers, and hydrogen fuel cells. We hope to work with them to apply this technology appropriately.

Design professionals need to communicate interactively to develop our own thinking and collaboration among design teams. It is essential to communicate effectively with a client if there is a consulting contract, or with a parent organization if the project is done by staff, or a funding agency for research projects, or with a constituency if it is an advocacy project. Review bodies also require presentations and submissions for planning approvals and building permits. Construction contractors and producers need documents for bidding and implementation.

Projects may also involve stakeholders, investors, or donors—all of whom will want to participate. There may be other parties to the project—various support groups or opposition groups, each having specific concerns. A key audience consists of the users of designs—the people who buy the product or use the facility. Design communication also is important for advertising and marketing. Facility managers and maintenance people are crucial to the long-term success of a design. They need good information to work with and can help improve the next generations of design. Design professionals need to communicate with all of these audiences

The design professions should be involved in educating the public— raising the level of design consciousness to help communities and cultures envision a better future. Design professionals are also involved in entertainment. Many of the content providers for movie, video, and computer games are design professionals who are reaching audiences that provide new markets for new media.

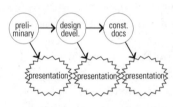

Presentation

Typically, design professionals make presentations at each stage of the project. We have to address the question of how we are going to package information for each presentation. How can we interact with our audiences—the review bodies, clients, users, and the market—incorporating their feedback? New media provide the opportunity to use information objects as the focus for communication as well as for collaboration. Using new media we can merge the models we are developing in our information environment into very effective presentations.

Many new possibilities arise. Design professionals can:

• Transfer presentations to videotape, as well as to paper.

• Produce CDs or DVDs (compact disks or digital video disks.)

• Set up web sites so the audiences can visit and preview design services and designs online.

Eventually we may be able to submit designs digitally so that planning and building jurisdictions could do their plan checks using parametric tools and checklists. Invite contractors and producers to use digital documents. Digital files, picked up from a web site, could ultimately be more useful for cost estimating and bidding than the paper documents typically provided. Digital documents also could be more useful for construction or production, improving project management. Some printing and manufacturing processes are now digital. Widespread use of digital documents in the construction industry could also make it easier to develop accurate as-built documents to use for future design modifications as well as for facilities and land management.

Guidelines for Design Communication With New Media

Using new media for online communication enables design professionals to:

• Connect with clients, organizations, and constituents to build relationships based on communication—not just fancy presentations.

• Expand the design team, engaging key parties to whatever extent they wish to be involved.

• Deliver what audiences need to know in interesting and useful ways. For example, we can include dialogue notes with attachments of shared models and key documents, letters, and memos focused on key points, short essays with "hot links" to other relevant sites, and illustrative storytelling techniques relating information in interesting and useful ways.

• Concentrate on core competencies and outsource complementing capabilities. (Generalists can team with specialists to create stronger designs. Professionals with different communication capabilities can work together in complementary ways.)

• Create collaborative teams that evolve online to meet the needs of each project. (The design communication skills of the team will become more refined, carrying over from one project to the next.)

• Engage stakeholders in realizing opportunities and involve their passion to bring projects to life.

• Share visions with customers—the real users of the design—and gather their insights.

• Gain continuous feedback to help guide the design as it emerges.

• Add as-built information, providing valuable documents for land facility or product management.

Information and Data Flow

Information flow diagrams are particularly helpful for understanding what pieces to present at different stages of the project. They often help identify the best way to package information for different audiences. Sometimes it is useful to present information flow diagrams as part of a presentation, so others can see where the information came from and how it is being used. We can look at what we produce at each stage of a project and recognize the information it contains. The object is to design approaches for easily transferring appropriate information from one stage of the project to another and disseminate it to key audiences.

Information flow diagram

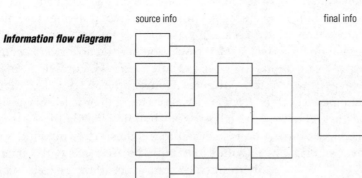

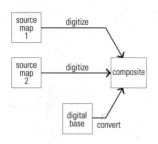

Compositing information

We need to develop clever strategies to acquire digital information. Besides using keyboards and pointing devices for digitizing, this involves working with scanners, digital cameras, digital video and audio recorders, and other instruments that gather digital information directly. We also can acquire useful digital information online and on CD-ROMs. Once we have digital information to work with, we may still need to translate these files into formats that work with different hardware and software. Standards are emerging that make these translations less of a problem. Translators exist to move files between different text as well as between different graphic formats. We need to examine the formats we are using and make sure we can move data from one application program to another.

Data flow diagrams help visualize these translations. There are often several strategies for translating files: Do we translate when saving files, or when opening files in a new program? Diagramming different strategies and testing them can determine which works the best. We can save time and effort by doing this before committing. A data flow diagram showing key transfers can also help our workgroups communicate.

When working creatively with computers, we do not deal with static information. We gather information and analyze and synthesize— adding insights. We evaluate designs and package them for presentation and implementation, continually transforming information throughout the project. Diagramming clever strategies not only can save time and effort but also clarify backup procedures at different stages of the project. This provides a safety net and enables us to go back to source information when pursuing alternatives.

Electronic media become very good integrators. We can take information from many different sources and pull it together using multimedia. Working on paper poses difficulties when drawings and maps are drawn at a variety of scales. By getting this information into a digital format (at its correct size), the question of scale no longer becomes an issue. Using digital tools, we can add overlays on drawings, maps or images as well as link attributes. For example, we can use a digital terrain model derived from an aerial photograph or field survey. This can be integrated with digital plans of a building if all of these plans have the same drawing unit (such as 1 foot equals 1 drawing unit). Often civil engineers work with 1 foot units and architects with 1 inch units, but this can be easily rectified by scaling one of the plans by a factor of 12 to make the drawing units the same.

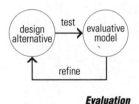

test

refine

Evaluation

Take care that accuracy doesn't become corrupted. It makes little sense to mix information that is obtained with great accuracy or high resolution with information that is approximate or of low resolution. The accuracy of the result is no better than the lowest level of resolution. Sometimes we need high levels of accuracy or high resolution, a lot of detail, or photorealism for the task at hand. But we do not always need this detail or precision. Pursuing perfection can be irrelevant and expensive. On the other hand, not having sufficient accuracy or detail can lead to expensive mistakes. When transferring information, we need to make careful judgments about the accuracy or level of detail necessary.

Typically we change a document when we work with it. Base information becomes combined with our additions. A template is no longer a generic form document once we add new information to it. If we want to use the template or base information again, we need to save it separately. This also makes it easier to update the base information and insert it or link it to drawing files. Consequently, we should develop strategies for layering or linking and saving base information as separate files.

Saving at stopping points

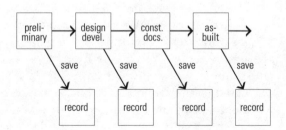

As we go through stages of the design process it is good to establish stopping points, saving documents at key stages of a project. By renaming the file we can save different versions at each stopping point. This provides ways of going back to base information and early versions of the design work if necessary. Careful file management can avoid getting different versions of the same file confused. A project team needs to be working on the latest set of documents.

Complex productions require a team with the expertise to deal with all the facets of the design communication. Work flow closely relates to information flow. Each activity results in a product. Each product has some information content. Consequently, information flow diagrams developing approaches to using media will help develop workflow diagrams. By relating work flow to information flow we can examine the best way to do projects. There is a tendency to go through the same pattern of workflow as was done on previous projects. Information and workflow diagrams can help visualize new possibilities to continuously improve what we do.

Diagramming information flow helps consultants and clients understand better ways to transfer information for the project. A good method is to proceed from source information to the finished product. Grouping different types of information (related by topics or types of information) can also help bring more clarity to the process.

In the research, or data-gathering, stage, we typically begin with some source data. This may be a primary source, such as a field survey for a project, or it may be a secondary source—such as a base map done for another purpose. People using the information should understand its source and accuracy regardless of whether it is primary or secondary.

The analytic stage involves interpretations. Each interpretation should have clear criteria. We can bring together many interpretations to create composites of information, deriving suitability composites (using positive criteria) or sensitivity composites (using negative criteria). For example, we can develop a model to analyze the best building sites based on interpretation of soils, slope, and vegetative cover. Compositing interpretations provide ways to integrate judgments of the information. It is usually helpful to present this to clients. Composites become a basis for synthesis.

Synthesis involves reacting to the information and drawing upon insights to give expression to the design. It is important to do this at each phase of the project. We typically start with a preliminary or schematic design presentation for a client. In design development we add more detailed information for review bodies. Only then do we proceed with implementation documents for contractors or producers. These presentations also require careful evaluation at each phase of the project. The information derived from the reviews needs to be incorporated into the design. Working digitally makes it easier to change the documents as the design evolves.

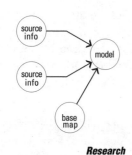

Research

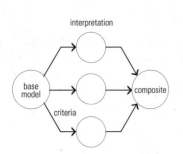

Analysis

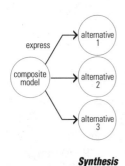

Synthesis

Mockups

Mocking up models and presentations makes them more tangible so we can relate more readily to them as information objects. Mockups for presentations may involve simple thumbnail sketches of pages or drawings, a block-out of an image, a script for a presentation, a treatment and storyboard for animation or video, or a cognitive map for linking hypermedia. Mockups conceptually show how the pieces of a model or presentation come together so we can decide what presentation software to use. They help visualize how to effectively use design models for design communication, improving coordination and reducing the time involved. Mockups also help us work concurrently on design and production.

mockup *A representation. Mockups show how the pieces of an object or the parts of a presentation come together. Enables visualization of a final product before producing it.*

The following types of software produce different presentations:

- Desktop publishing software—printed pages

- Illustration software—posters and presentation boards

- CADD—drawings and 3D models

- GIS software—maps and databases of spatial information

- Presentation software—slide and video projections

- Web publishing—web pages with links

- Multimedia software—videos as well as interactive multimedia online or on CD-ROM

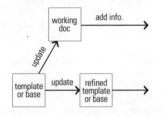

Transferring bases

Quick thumbnail sketches on paper are often all that is needed to begin. Once a format is established, it may be easier to work directly on a computer. Presentation software, desktop publishing software, web publishing, and desktop video or multimedia software all provide ways to mock up presentations as we develop them. We can work with presentations interactively by either adjusting the layout or going back and refining the pieces so they are more appropriate for the presentation.

Multimedia

Multimedia programs help produce presentations integrating text, graphics, sound, and video, providing new possibilities for human interaction. The audience may view a multimedia presentation passively, as people do watching videos and films. Or the audience may explore multimedia presentations interactively online or using laser disks. This enables people to play segments of full-motion video related to topics of interest to them.

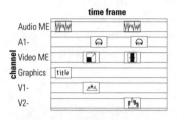

Multimedia storyboard

There are new opportunities to engage producers and construction contractors using digital design communication for computer-aided manufacturing and other seamless methods for implementing designs. We can transfer files to them online. Or we can put large design documents on disk or CD, providing documents that are more compact and less costly to produce and distribute than those traditionally printed out on paper.

There is a difference between being a viewer and a director, just as there is a difference between being a reader and an author. Multimedia software is geared to directors—people who can tell a story by creatively assembling multimedia information.

Visual literacy should involve learning to perceive, develop, and express ideas and information using more than one medium at a time. Developing greater visual literacy makes it easier to establish cyberspace work environments where people view, interact with, and express ideas and information using multimedia computing. Workgroups can achieve new levels of collaboration. The type of close, personal interaction people experience working with physical objects in reality also becomes possible when working with imaginary objects in virtual reality.

The evolution of media is also changing the way in which we pursue design endeavors. Centuries ago landscape architects, such as Capability Brown, would work directly with the land. They worked with the actual reality (consisting of the landscape and social context) and the vision (or mental image they used to develop their design). It was difficult to communicate the vision without implementing the design. Implementation of the actual landscape took considerable effort and resources. So landscape architects most often worked for rich patrons who had the resources to construct these visions. The landscape architects and master gardeners worked the design out on the site.

As drafting and delineation evolved, designers worked more plans out on paper. Paper drawings became a design and presentation medium for communicating with clients, users, review bodies, and construction workers. However, this is changing. Using computers, landscape architects can, in effect, create artificial realities with electronic media. The actual site can be photographed from an airplane or satellite. From this, a digital survey can provide a three-dimensional model of the landform. The planning and design team can work with this model in a computer, mapping invisible attributes such as geology, buried utilities, and zoning. The planning and design team can do site analysis and interpretation using this information, understanding the land as a living ecosystem. The landscape architect can use this information to model the design. Other design professionals—such as architects and engineers—can share these and other models as a focus for collaboration. The stakeholders and users of the environment can be involved in helping to improve the design. Of course, design professionals also need to relate to the real site as well, bringing meaning to the multimedia models they develop.

Recognizing and representing realities

reality relates to virtual reality

physical space relates to media space

New media enable people to create a new reality. As professional design practices continue to evolve, it is likely that we will do more implementation directly from the models of the artificial reality. Product designers can link the manufacturing of products directly to models created with computers, without doing traditional presentation and construction drawings. They can develop implementation procedures for manufacturing concurrently as they develop the design. This can result in better coordination and quicker implementation. Other disciplines have similar opportunities to model objects in cyberspace and transfer them electronically for implementation. Building architects, landscape architects, interior architects, and engineers can work interactively with digital models they share online, improving the communication and coordination of extensive teams needed to design and implement the built environment. Graphic designers can mock up the layout of a report electronically while authors and artists work on the manuscript and graphics. Movie producers can weave digital special effects into feature films and TV productions. Design professionals can help create computer and video games that take place in virtual reality. New art forms can emerge in cyberspace for people to visit online and experience interactively.

Value
While some aspects of design (acquiring base information and using standard details and specifications) may be automated, the nature of original work limits this potential. The real value of digital tools is in enhancing communication to enable better collaboration and coordination so teams can work more concurrently. Interactively using digital design models in this way, design professionals have already made dramatic gains in the speed with which new products can be developed.

Online design communication can be coupled with e-business to manage enterprises. The emerging business models for virtual offices present competitive advantages. Online design communication can be coupled with the e-commerce that is rapidly growing. Coupling design with commerce makes it more possible to do mass customization, meeting the needs of individuals and niche markets. This creates new opportunities for design professionals to produce and market what they design online. No longer do design professions need to be subjugated to business and marketing as usual. Designers may be able to gain more control over branding. More of the revenue generated can be focused on design—optimizing and improving design quality. This can involve greater efficiencies and fewer environmental impacts, resulting in overall better value.

The content that design professionals provide also has value for education, promotion, product maintenance, and management. Doing more with the content takes more time; however, by producing more value, design professionals can earn more compensation.

The information objects design professionals create using new media may become collectible—in the same way that traditional art objects—photos, drawings, paintings, and sculpture—are collectible. In addition, virtual reality models could be incorporated into entertainment and possibly even computer or video games. Writing, composing music, or drawing and painting each are art forms with long traditions. People specialize—as writers, poets, composers, painters, sculptors, architects, or product designers—to master the knowledge required of a discipline and the complexity of the media they use. The tools available today are making it easier to work in multimedia, providing us with the potential for fuller expression. Artists and other design professionals can express their creative thinking using multimedia. From this, new art forms and products can emerge.

Better design communication can shorten the path from a concept in one person's mind to comprehension of that concept in another person's mind. New media provide a vehicle for collaboration and implementation of design ideas. If people can understand new concepts more quickly and thoroughly, they can draw upon more of their mental capacity. Applying intellectual capital involves not only using knowledge people have acquired; it also involves their capacity to learn and respond to new situations creatively, generating innovative ideas. It involves the capacity to bring expression to an artistic level that resonates with the human spirit. Online communication is becoming part of the web of life.

Summary

Methods for Communicating Designs Interactively

1. **Identify the audience(s).** Think through how to package information appropriately for each audience. Use multimedia to open more channels for communicating the message. Use new media to put together a variety of presentations. Make the presentation more interactive, engaging the audience in providing feedback that can add value to the design. Incorporate that value into the next design iteration.

2. **Diagram information flow.** Think through how to transfer information from sources to the final products, adding value along the way. Look for base information you can add value to.

3. **Mock up how to package information for presentations.** Find effective ways to interact with your audience. Decide what presentation software to use. Interactively use digital design models in the mockups to get a sense of how everything fits together for the presentation.

4. **Diagram data flow to work out hand-offs of digital information.** Test these transfers to make sure the files are compatible on the hardware and software the team is using. Arrange for handoffs of digital information online—attached to e-mail, transferred through a web site, on disk in meetings, or overnight deliveries of CDs with large files. Carefully consider the production schedule, making sure you can produce what you want on time.

5. **Use digital design models to develop the necessary implementation documents.** Some design disciplines—such as publishing and product design—can hand off digital files for production. Other design disciplines, such as architecture, and landscape architecture, typically print out traditional construction documents. Look for opportunities to work digitally with production and construction teams. You may do this through design/build or by collaborating with contractors or producers to implement digital design documents.

6. **Derive residual value from the content you are producing.** Where appropriate, package information for publication or promotion. Provide as-built documents for facilities and land management.

Activities
Design Digital Communication

The challenge is to express design ideas electronically, sharing base information with consultants, and communicating design concepts with collaborators as well as clients and local review bodies. Effective design communication is essential for implementation. Implementation documents done in digital formats can be used for construction bidding, administration, and management.

Select a project to present digitally. This could be the project you started in a digital design journal and developed using digital models. Think how to interactively use new media for design communication. Maybe you already have been sharing parts of your design journal with your project team or used models to explain the analysis and preliminary design. Who are your audiences? What information do they need? How can you provide it most effectively? And how can you engage them to get their input?

Mock up implementation documents and diagram the information, data, and work flows involved. This will help you, and your team, visualize both product and the process. Mockups show the product. Flow diagrams show the process. This facilitates figuring out who is doing what, and when it needs to be done. Use mockups and flow diagrams to manage the project effectively.

Each team member should have a clear picture of what he or she is producing and how it fits into the overall package. When working digitally, it is good to identify hand-offs including what the file format will be and how to transfer the file. On disk? On CD-ROM? Online? Test the hand-offs to make sure file formats are compatible with the hardware and software your team is using. Evaluate file sizes and timing when considering the media to use for transfers. Coordinate who is producing what, and when. Much of this coordination can be done online, particularly if the project involves a virtual office using online services to engage people all over the world.

How will you interact with your audiences—the review bodies, clients, users, and the market? You will need to make a series of progress reviews and incorporate their feedback. Look for opportunities to use digital design models throughout the production/construction stage as well as for post-production and land/facility or product management.

Modeled and rendered by Jason Jamerson and Scott Davidson in AutoCAD 14 using AcuRender 3.

Case Studies

"How to Do It"

PART I Methods

Design teams—including managers, design professionals, and staff—need ways of working effectively. This part of the book offers methods to use new media and digital tools creatively and productively for design communication in professional practice. It can be a helpful text in courses on design methods and new media for design communication, as well as computer courses that go beyond application training.

PART II Case Studies

Good examples demonstrate how progressive design professionals are using new media and applying digital tools in their practices. The case studies examine a range of design practices—writing and graphic design, creating multimedia and special effects, landscape architecture, architecture, and planning. The case studies explore what works in office environments as well as in virtual offices and electronic studios using online collaboration.

PART III Strategies

Once we invest and commit to using new media and digital tools, it is important to find ways to get the most out of them. This part of the book is devoted to realizing the potential inherent in information technology. It will provide valuable help as your design team addresses change.

APPENDICES

6

Publishing: On Paper, Online

If our team were to collaborate online to create this book and CD, we would have more freedom over when and where we worked. Online collaboration would also make it easier to produce the book in less than nine months, which is what our contract called for.

I launched this project with this among other goals in mind. Certainly, producing a quality book in a short time would require drawing not only upon my own years of professional experience as an educator and design professional, but also that of others. A sabbatical from Cal Poly Pomona provided time to focus on writing. I selected a team with the capabilities needed to produce the book in a short timeframe.

Maintaining a balanced life and having freedom to travel were important to me while working on this project. Digital tools provided access to information, contributors, production staff, reviewers, and editors. Moving documents electronically saved time. By collaborating online with my team, I was able to spend more time with my family. A notebook computer and modem offered the freedom to work wherever and whenever I wanted to.

The Team

The team included staff from my office to help produce the graphics and deal with the administrative chores. Graphic designers laid out the book. There was a CD designer as well. The team also included many people who reviewed and edited the material.

How do you put together and coordinate an effort like this? You need capable people creatively using digital tools.

As the author, I like to write with a notebook computer at my cabin at Big Bear Lake, California. There is no phone and few interruptions. I can wake up with a clear head and write without being drawn into the crisis of the day, which can be all-consuming. When I need a break, I enjoy a refreshing walk in the woods. However, by hiding out in a retreat (or a study), authors can lose touch with their families. Fortunately, I like to write early in the morning when most people in my family are still sleeping. There is a need to spend time with people you love. They are a hidden part of any team.

My wife Carla had surfed the net booking hotels from Holland, to France and Italy, where we would spend time with our daughter at a country house in Tuscany, near Barberino. Kirsten, who lives in Italy with her husband Francesco, also booked us a room at an inn near

68

Ortisei in the Italian Alps, and a hotel room on the island of Elba. I needed to find a way to balance being with them—enjoying this summer journey—with my commitment to this project that I had contracted to do—a journey of a different sort.

I also wanted to spend some time with my father who lived in New York State. He was turning 90, but still provided me with newspaper clippings from *Wall Street Journal* articles dealing with computers. Unfortunately he died at the end of the summer, just before I was able to visit him. After the funeral, our extended family had to empty the family house, sell it, and settle the estate. I took along a notebook computer and stayed in touch with my office thousands of miles away, but was unable to do much writing on this book for almost a month.

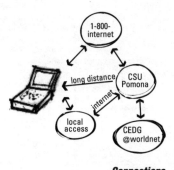

This delay meant that I did not complete the manuscript during my fall quarter sabbatical. Continuing to write—with a full teaching schedule, extensive e-mail correspondence, and my regular office practice—resulted in a repetitive motion problem. Using a keyboard and mouse became painful because tendons in my arms and back were inflamed. My fingers became numb. How could I advocate using digital tools, if their use resulted in injury? I resolved to avoid excess use, and pace my writing sessions by stretching frequently. For the rest of my writing, I shifted from a notebook computer to using an ergonomic keyboard while sitting in a supportive chair, properly set up at a workstation. Voice recognition ultimately saved keystrokes and time. Acupuncture and yoga also helped me heal by the end of spring.

Connections

Fortunately, in the summer, I was able to finish this book even though we had passed our deadline. Jared Ikeda, my good colleague, helped pull together examples for the CD and find sites for the Links Library. Professional staff helping with this project at CEDG included Daniel Cressy and Jonathan Hartman. My son Erik also helped with graphic production and testing the CD. Judy Casanova was a superb editor. Brooks Cavin, my partner, offered comments and support. Our professional office does architecture, landscape architecture, and planning, focusing on sustainable regenerative design, which many people call "green architecture." We design with a variety of digital tools, including Photoshop, Adobe Illustrator, AutoCAD and Form Z.

Clinton Wade and Mary Stoddard, Clint's associate, were responsible for the book's design and production. They have considerable design background and expertise with QuarkXPress and other tools of the graphic design profession. Elyse Chapman created the CD using QuarkXPress and Adobe Acrobat. She came to the project with considerable design and technical background and expertise with Adobe Acrobat.

Our Shared Environment

Our information environment developed in four domains: the project team's computers linked by e-mail, an FTP site, a web site, as well as Zip disks and CDs that we used to transmit larger files and the finished product.

E-mail was our main mode of collaboration. It was the primary way in which I maintained contact with the publisher as well as with those who contributed content for the case studies. Even though the rest of the team lived nearby, e-mail proved to be a convenient mode of collaboration. I would send manuscript to Judy, my editor, as attached files. She would insert edits and return the attachments via e-mail. Then I could refine the manuscript and hand it off, along with the

Shared information environment

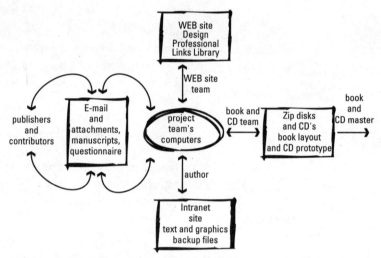

graphics Erik produced, to Mary who put the book together. It was important to establish a consistent file naming system that everyone worked with. Each person was responsible for backing up his or her own work. We all needed to maintain virus protection.

Towards the end of the project, Warren Roberts and I established an FTP site on the Intranet at Cal Poly Pomona. Since Jared, Warren, and I teach at Cal Poly, we could share access to this site. I also wanted to backup files at another location, to protect them in the event anything happen to the files in our personal and office computers.

Jared set up a web site for the Design Professionals Links Library where we collected the URLs and compiled annotations describing them. Since this was a public site, we could invite others to visit online, try out the links, and suggest more sites.

Zip disks and CDs continued to be primary modes for backing up files and also for transferring work. The Quark files for the book and the examples that we integrated into the CD were particularly large and would take a long time to transfer via modem, so we transferred these files using disks. We put the final digital files for the book on a Zip disk and sent it to McGraw-Hill in New York. We produced a master for the CD accompanying the book. Using the CD-R in my professional office, we also archived the project documents.

Large organizations with Intranet sites and fiber optic local area network connections can set up shared information environments for project teams. Increasing telecommunication speeds will make it practical for more design professionals to set up share information environments through service providers and transmit documents and drawings, as well as images and 3D models, using high speed internet connections.

Workflow

In organizing and managing a project like this, it is important to have an established workflow to move information from the sources to the final product with as little effort as necessary.

I write in several ways. Often I just brainstorm, jotting down key words and drawing on a notepad, sketch book, the back of an envelope, napkin, or any thing else I can find when inspiration strikes me. To record insights, I like using voice recognition and a word processor to follow insights as they emerge. Then there is the process of editing—bringing these insights into clear focus and telling the story. Of course, these ideas need to fit into the framework of the book. Sometimes the most difficult task is selecting which ideas to include. Another way of writing is to expand an outline—deductively filling in details of this larger picture by logically drawing upon source information. This is done by dictating with notes in hand and expanding an outline using a word processing program.

Work flow diagram

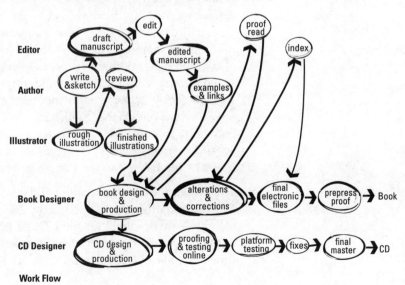

Work Flow

I did some of the interviews for the case studies by taking notes and taping portions of the conversation to help me review what was said. Some of the interviews I did online by asking key questions through e-mail correspondence. I e-mailed the case study drafts to the subjects, giving them an opportunity to review and add to what was being said. So we collaborated interactively to get the essential ideas across.

I like to draw when I write. Sometimes ideas start as sketches that I then write about; sometimes ideas emerge as key words that sketches help me explore. Graphic visualization is a wonderful language for ideation. Sketches can range from being abstract and fuzzy to being concrete and descriptive. Moving between the abstract and concrete helps me understand ideas and clarify them for others.

I find traveling can be a stimulus for writing. It induces fresh insights, gets me out of my routine and away from other obligations and responsibilities. Traveling can also provide time to reflect. The new portable digital tools, with modem connections, offer the means to collect thoughts anywhere and disseminate them online. Collaborating with a team helps me get what I write into a publishable form to share with others.

My editor in New York received the draft manuscript via e-mail for review. I handed off rough sketches to my graphic production staff in California. Sometimes I faxed them or attached them to e-mail; sometimes I produced digital primitives of drawings they refined using the digital tools in Adobe Illustrator and Photoshop.

Once we had refined text and graphics, we transferred the files to our shared information environment, so the book design and layout team could pick them up and work with them. At the beginning of the project, we worked out the design of a sample chapter of the book, using Quark XPress, so we knew how many manuscript pages and how much space there was for graphics in the format we were using. We also tested this production process to make sure it worked. This provided a framework for the text and graphics. The formatted document was then copy-edited, and the necessary corrections made to the final files. We sent files to the publisher on a Zip disk along with verifying hard copy to fulfill our contract for the book, although with high-speed Internet connections, these files could have been transferred online. The publisher handed this off to a printer to produce the book.

To produce the CD, we adapted and converted portions of the Quark XPress formatted book—such as the goals, chapter summaries, and activities—to Acrobat PDF files. These, along with the glossary, bibliography, and overview chart, became key components. At the same time, we collected more graphic material that became the examples

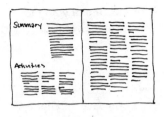

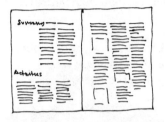

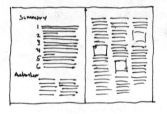

Progressive layout sketches for Summary and Activities spreads. Preliminary pencil concepts and first computer-generated layout, which was refined for the final design.

you find on the CD linked to each activity. We obtained much of this material from design professionals working online and from computer shows such as AEC Systems and SIGGRAPH. We refined this material to read well on the screen, and we worked from a cognitive map developed for the CD, linking needs to goals to procedures to activities to examples and to online resources so that the user could navigate this .resource information. At the same time we set up a web site to establish the Design Professional Web Links Library that we keep updated in the College of Environmental Design at Cal Poly Pomona. Jared Ikeda, Kyle Brown, Warren Roberts, and others helped me establish the Design Professional Links Library. Having material on line made it relatively easy to invite people to try it out. The challenge became to get all the CD links working together and testing them on Windows and Macintosh platforms. Once we had the CD tested, we burned the master and sent it to McGraw-Hill for production. They sent the CD out for duplication and distribution with the book.

You may be reading this book to learn methods and strategies for designing with new media by creatively using digital tools. Or, you may want to use the CD to quickly access only the information you are interested in and see multimedia examples. The CD also serves as a resource guide to using new media. McGraw-Hill may make portions of the CD available online as an overview of what we offer in this book and CD for marketing purposes.

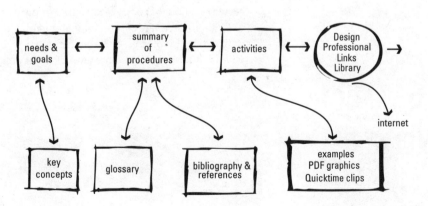

Cognitive map of CD

Summary Guidelines for Publishing

1. Be clear about your goals so that you can plan production of your publication. Target your audience and determine the best way to deliver your message to it.

2. Identify the tasks involved, establish your team, and select the digital tools you will use. Remember: there is a learning curve when using new tools, but digital tools can help you do more, faster and better. You have the choice of investing in the tools and learning to use them, or hiring those with the tools and expertise to be part of your team.

3. Determine the optimal workflow necessary to get your work done effectively. Diagram workflow to develop procedures and help manage your project. Record what works and doesn't work so you will be able to develop better procedures in the future

4. Establish your shared information environment by setting up file-naming conventions, format standards, and procedures for handling the flow of information. It can be helpful to diagram your information flow, focusing on the best way to transfer files between software and hardware platforms, and making sure this works before you commit.

5. Develop a model or prototype of your publication or CD. Make sure that you can transfer all the necessary files, and that the publication's format works well for your audience. You can do this by testing sample chapters of a book or segments of a CD.

6. Produce the document using your online information environment. You can also use this format to have people review and test it online before you release it for publication.

7. Hand off your document for publication. This may involve submitting your files to a publisher—many are now picking up digital files online or accepting them on disk. This may involve getting your final document files to a service bureau that does the printing or burns the production master CD from a digital master; larger professional organizations will do this in-house. Or, it may simply involve posting the document to a web site. Make sure you are satisfied with the quality before you release it online.

Activities
Publish

Design professionals often need to produce marketing information, proposals, reports, and sometimes, publications. The challenge is to learn how to do this using new media. The new media enable you to produce high-quality documents, link up with publishers and service bureaus that do printing and even publish online.

Select a project you are working on that would provide an opportunity to explore new media for publishing on paper or online. Use it as a pilot study to help change your professional practice, enabling you to enhance your capabilities and improve the quality of what you do. Select tools that will be appropriate for your task. Either build into this project the time you will need to learn the necessary tools—word processing, graphic production and imaging, graphic design, desktop publishing, and web publishing software—or build a team that includes people with this expertise.

Organize your team using approaches similar to those described in this chapter. You will need to visualize your workflow and determine who will be responsible for each task. If you are working in-house, this will help you manage your project. If you are subcontracting portions of this project, this will help you establish and coordinate those contracts.

Establish your work environment. This may be an office where you have the physical space to set up for the project. This also may be online where you can create an information environment that each team member can access. Communicating online can make it easier to link people working in different places on different tasks.

Develop an effective file management system. Whether you are working together in an office, or a studio, or online, it will be important to have a file management system that everyone understands and works with so that you can all manage digital files effectively, keeping track of the latest versions.

Use this case study and the summary provided in this chapter as guidelines to help you proceed. The CD contains some of the working files created to produce this book. You can explore these files if you want to get a sense of the organization of our information environment, and the file management system we used. You will also find some of the correspondence for the project.

7

Design Visualization: Realizing a Better Future

Design professionals can visualize ideas that may lead to a better future. The challenge is to use new media creatively for design communication and implement these ideas. I believe we have a social responsibility to address issues facing society and provide vision to deal with them.

Having lived in the Los Angeles Basin for 30 years, I am deeply concerned about cleaning up the air. I find myself visualizing ways to prevent the air pollution that damages public health and the environment. Many years ago, this passion caused me to write about the "carrying capacity of the Los Angeles Basin" and fill sketchbooks with ideas for passive solar buildings and better bicycles. Today, it is a focus for my commitment to the Cal Poly Pomona Center for Regenerative Studies (CRS), founded by my good friend and close colleague, John T. Lyle.

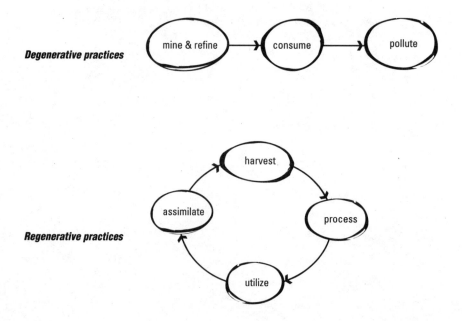

Degenerative practices

Regenerative practices

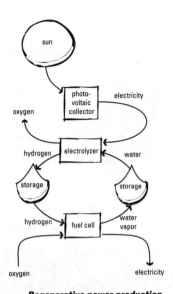

sun

photo-voltaic collector — electricity

oxygen

hydrogen — electrolyzer — water

storage — storage

hydrogen — fuel cell — water vapor

oxygen — electricity

Regenerative power production

Clean Air Park

In 1997 when Bob Zweig, founder of the advocacy group Clean Air Now (CAN), asked me if I was interested in moving their Solar Hydrogen Project to the CRS, I became excited about the prospect. CAN had developed a $2.5 million demonstration facility at the Xerox site in El Segundo, with funding from the Federal Department of Energy and other entities and was offering it to the Cal Poly Pomona CRS. This facility demonstrated how photovoltaic panels could convert sunlight into electricity to electrolyze water, producing hydrogen to fuel several pickup trucks with internal combustion engines converted to run on this fuel. Emissions from these trucks were mostly water vapor. Fuel cell technology could make this demonstration even cleaner and more efficient. It seemed to me our Center for Regenerative Studies should be involved in this type of demonstration. I needed to convince the university administration to help obtain funds from the South Coast Air Quality Management Control District (SC AQMD), the California Energy Commission (CEC), and the U. S. Department of Energy (DOE) to move this facility from El Segundo to the Cal Poly campus.

Working with two teams of students (the first a group of undergraduates taking the Advanced Landscape Design Course that I taught with Jared Ikeda, and the second, graduate students in the Landscape Technology Course that I taught with Warren Roberts), we developed a vision of how the CAN Solar Hydrogen demonstration facility could be integrated into the CRS. We called this the Clean Air Park Proposal. We added a wind generator to demonstrate another way to harvest renewable energy to store as hydrogen for use in campus vehicles or fuel cells for electronic media and the CRS buildings when the sun doesn't shine. We suggested having a clean fuel station on campus that could earn revenue by retailing reformulated gas, methane, and hydrogen for zero (and low) emission vehicles coming to campus or traveling between Los Angeles and Las Vegas on the "Hydrogen Corridor" advocated by the Hydrogen Industries Council and the DOE. We proposed to demonstrate how to reform methane from the landfill (that is part of the Cal Poly Pomona Landlab) to produce hydrogen. This involved a solar reformer being developed by Jerry Cole and others at EER (Energy and Environmental Research Corporation.) In addition, we proposed producing methane from a digester that assimilated green matter from the gardens at the CRS and Agriscapes. We even envisioned growing hydrogen-producing bacteria and algae. Using straw-bale construction for these facilities, we could demonstrate ways to sequester carbon that typically goes into the atmosphere from agricultural burning.

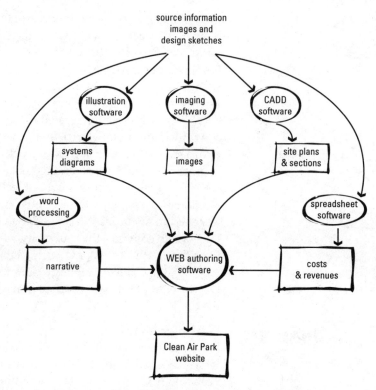

Information flow of student efforts

The student team graphically developed systems models conceptualizing how all of this could work. Using CADD, we developed a site plan showing how this could fit into the land at the CRS and spreadsheet models projecting costs and revenues. In addition, we developed images to show people what this facility could look like at the CRS. My students posted their work on the World Wide Web to share it with the University administration and prospective funding sources. You can still visit the Clean Air Park website through the Cal Poly Pomona Landscape Architecture website. Portions of this website are on the CD accompanying this book.

Unfortunately, the Campus Planning Committee was not able to consider the Clean Air Park proposal. The university administration was preoccupied with a proposed golf course surrounding the CRS. There was a lack of vision regarding the need to research and demonstrate renewable energy, and an unfounded fear of hydrogen. Although unable to overcome these obstacles, the Clean Air Park Proposal did provide a wonderful educational experience, enabling students to visualize designs and use digital tools to develop and communicate them with new media. A Cal Poly Pomona Research and

Creative Activity Award enabled me to put together recommendations for integrating Living Energy Systems into what is now renamed the John T. Lyle Center for Regenerative Studies, relating it to the Phase III Plan John Lyle did before he died.

Living Energy Systems

As the Clean Air Park Proposal stalled, Mark Sorensen, the president of Geographic Planning Collaborative (GPC) contacted me about doing a project for an orphanage in Vietnam. One of his clients, Mrs. Matsuda, of the Japanese Housing Organization, is a benefactor of the Bien Hoa Orphanage. She wanted to help this orphanage develop an energy center using a regenerative infrastructure similar to what we had been working on at the Lyle Center for Regenerative Studies. Our architectural, landscape architectural, and planning firm—the Claremont Environmental Design Group (CEDG)—had worked previously with Sorensen on projects for Mrs. Matsuda in Japan as part of an international team planning energy-efficient communities. Using online collaboration to organize those projects when he was at Environmental Systems Research Institute (ESRI), Sorensen managed large teams involving planning and design professionals on different continents working around the clock with digital tools such as GIS, imaging, CADD, as well as publishing and telecommunication software.

Welcoming this opportunity to continue working on "living energy systems," we formed a team involving my professional firm, CEDG, as well as technical advisors who had worked with me at the Lyle Center for Regenerative Studies. The collaborative team for this humanitarian endeavor included renewable energy advocates, clean air advocates, landscape architects, architects, engineers, manufacturers and distributors of renewable energy technology, as well as biological wastewater treatment and permaculture experts.

This project began just as I was leaving for Europe, so it was important to set up online collaboration. Sorensen, whose office is at Running Springs in the San Bernardino Mountains of California, kept in touch with the client in Tokyo. The CEDG office is in Claremont at the base of the San Gabriel Mountains and our technical advisers were located throughout California. We communicated quite simply through e-mail, some phone calls, and a few faxes. We had an initial meeting at CEDG to examine base information (photos and base map that Sorensen had obtained visiting the orphanage), discuss problems and opportunities, and establish design directions. Then I departed for Europe but continued working and collaborating online to develop regenerative design concepts that were appropriate for the orphanage in Vietnam. Because I was to be gone for two months, I had planned to work on this book and office projects using a notebook computer. I found it intellectually stimulating to think about "living energy systems" while staying at a country house in Tuscany and visiting the island of Elba where I could experience civilization that has endured for centuries.

When I returned, we had another meeting for the Bien Hoa Orphanage project at CEDG—this time also involving the technical advisers to help sort out the concepts and develop the plan. Daniel Cressy and others at CEDG had digitized the base map and, with Brooks Cavin, began to develop the master plan using Autocad. We also developed sections to examine profiles of the plan. Using Adobe Illustrator, Cressy worked up the systems diagrams I had sketched. Jens Lerback, architect and freelance delineator, used the digital master plan and developed a three-dimensional wire-frame model to refine the form, select views, and block out the renderings. After the plan and renderings were colored, a service bureau with a large-format scanner transferred them into a digital format. I put together the narrative and selected images from the NREL website to show the technology we were using. Sorensen and I laid out the presentation, and he put it together using Pagemaker. After plotting and mounting the images on foam core, the boards were flown to Tokyo for Mrs. Matsuda to take to a meeting in Vietnam where she presented them to officials from the orphanage and regional government. Using Adobe Acrobat we have compiled the presentation into a .PDF file which is included on the CD accompanying this book.

The orphanage presently consumes tons of firewood just for cooking and heating water. Sanitation is marginal, and there is little hope for expanding the orphanage given the present infrastructure.

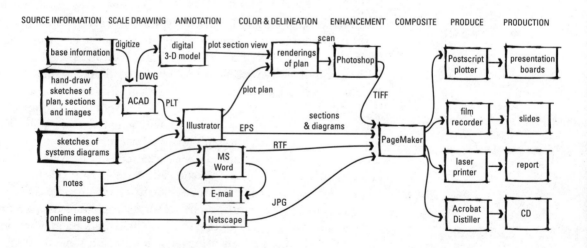

Information flow for the "Living Energy Systems" project

The purpose of this master plan is to:

• Provide a sustainable infrastructure to expand this orphanage in Vietnam, enabling it to accommodate more children.

• Demonstrate ways to integrate distributed power production that would convert locally available renewable energy into electricity at the point of use.

• Conceive "living energy systems" that, in addition to producing electricity, could provide fresh food and clean drinking water, heat for cooking and hot water, process sewage, and reclaim water, as well as nutrients for permaculture.

• Plan for a technical school to provide training in the ongoing use and maintenance of solar technologies and permaculture.

Using digital tools, this plan models possibilities for assimilating distributed power production into development in ways that reduce impacts on ecosystems upon which the world depends for life. Regenerative design can help sustain bio-geo-chemical cycles, and avoid pollution. This plan shows how integrated living energy systems could work in a community, building complex, and single-family dwelling in the context of Vietnam. Distributed power production could become an integral part of living energy systems throughout the world.

The intent of this plan is to enable the regional government and orphanage near New Saigon to obtain funding from international humanitarian, environmental, and economic development organizations to improve living conditions. The implementation of this project would help people throughout the world visualize ways to reduce greenhouse gases, avoid dependence upon fossil fuels, and stimulate economic development using local resources. The role of the design professionals involved in this project has been to provide regenerative designs that integrate energy, food production, water, sewage treatment, and compost systems into the site's ecological context.

Living energy systems demonstrate environmentally responsible ways to produce both electricity and food. This project, which calls for shifting to solar and hydrogen energy, would stimulate the development of appropriate energy technologies. Distributed power production (and the knowledge to make it work) could enable more effective harvesting and use of renewable energy, thus addressing global greenhouse problems and the eventual depletion of fossil fuels as well as the need to provide healthy food and clean water on a sustainable basis.

The American Society of Landscape Architects selected the Bien Hoa Orphanage Living Energy System Project for a 1999 National Merit Award for Research and Planning. The Architects for Social Responsibility-Boston Society of Architects, in conjunction with the Environmental Committee of the New York AIA, also recognized this project with a 1999 Sustainable Design Award for Planning Vision.

The Kyoto Mandate

Students in the Cal Poly Pomona landscape architecture graduate program, taking the "Landscape Design and Natural Processes" course I taught with Kyle Brown, also addressed local and global air quality problems. We asked the students to visualize ways of going beyond emission controls while implementing the Kyoto Mandate to "reduce greenhouse gases 7 percent below 1990 levels." Their projects demonstrate ways to improve the quality of life in the Los Angeles region and around the globe.

Many of these students had not had the opportunity to do much with digital tools. However, they quickly developed capabilities to communicate their design ideas through PowerPoint presentations and web pages. Each team compiled information, developing design journals they could transform into presentations using digital tools. Their design communication evolved as they developed their ideas. Some of the students were able to develop location models using ArcView GIS software. Working with air photos they scanned, or obtained digitally from the Web, students used Photoshop to develop renderings of their plans. They used graphic programs to diagram material and energy flows and developed spreadsheet models to evaluate feasibility. Through their projects, they learned to design with digital tools.

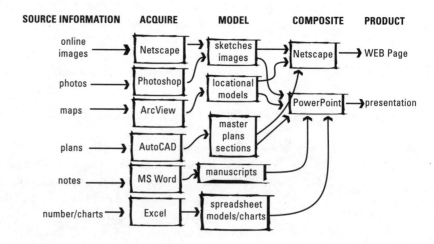

Information flow typical of LA 602 Sstudent projects (Each team worked with variations of these basic processes.)

The class pursued three strategies to demonstrate how the Kyoto Mandate could be implemented through regenerative landscape architectural design: One strategy created more efficient patterns of settlement and transportation reducing dependency on fossil fuels. A second strategy demonstrated ways to make use of waste (including greenhouse gases.) A third strategy enhanced the "green lung" capacity of open space to assimilate air pollution.

The students selected case study projects from the South Coast Air Quality Management District (SCAQMD) and assumed roles to apply their design expertise to air quality issues.

In "Claremont's Village Expansion Project" Jay Brown, Jeffry Stevens, and Rochelle Tortorete, assumed the role of a land developer and explored ways to include more energy-efficient housing into the new urbanism being planned for the Claremont Village. Their development proposal also included a park that could serve as a green lung with trees sustained by absorbing storm water and using gray water from the housing development. A plastered straw bale sound wall along the railroad tracks could sequester carbon from straw often burned in agricultural fields. Photovoltaic cells mounted on the sound wall could provide electricity for the park as well as for the power grid, thus demonstrating how to integrate this clean renewable energy source into development. These students presented their work to members of the Claremont City Council and Planning Commission.

Through the "Greening of Pilgrim Place," Brad Harris assumed a consulting role to demonstrate how a retirement community in Claremont could use reclaimed water to enhance its urban forest, improve energy efficiency in homes, and make greater use of electric vehicles. All of this would improve air quality while reducing long-term operating costs.

Shelley Harter, Aerin Martin, and Mina Nakanishi, assumed a consulting role to the AQMD, as the "S.A.M. Institute" for setting up public education to improve air quality in homes and in communities. This project proposed a certification program that could enhance property value by improving air quality.

In the "Urban Green Infill Project," Kibum Sung and Miki Yanai Hernandez assumed a consulting role to Cal Trans with participation of the AQMD. Using the public right-of-way at the intersection of the 10 and 57 freeways, they demonstrated ways to create a freeway landscape that would filter particulates and absorb emissions, thus enhancing air quality along the freeway. Their project showed ways to incorporate more bioremediation into the landscape as well as on sound walls and bridges already provided along freeways in urban areas. Gas tax revenues could provide funds for this mitigation of emissions. Japan and places in Europe are beginning to do this to a greater extent than is happening in California.

Through "Dungro" Doug Delgado and Michael McGowan assumed an entrepreneurial role, setting up a venture that would use manure to produce electricity, fertilizer, and other useful products. They selected a site in Chino near the dairy farms and designed a facility that would demonstrate the processing of manure while producing package treatment plants to sell to large-scale dairy farmers all over the world. Revenue from using manure to produce electricity and fertilizer could pay for package treatment plants, which would significantly reduce methane emissions from dairy farms and thereby help improve the global greenhouse gas situation.

In the "Chino Creek Watershed Project," Haiyan Ye and Adam Kringel assumed the role of Army Corps of Engineers staff. Their challenge was to expand the mission of the Army Corps to combine both watershed and airshed management responsibilities. Riparian areas have the potential to sustain considerable vegetation that can absorb carbon dioxide, produce oxygen, and filter particulates, while at the same time protecting urban areas from floods, percolating water into groundwater basins, and enhancing wildlife habitat. This project demonstrates how this could be done along Chino Creek.

Some examples of these student projects are on the CD accompanying this book. You also can view these projects and others at the Cal Poly Pomona Department of Landscape Architecture website.

Beyond New Urbanism

The California Central Valley is expected to have an additional ten million residents by the year 2040. The California Council of the American Institute of Architects, along with other entities, hosted a design competition to generate ideas for housing these people. Realizing that many of the ideas we had been working with were applicable to the Central Valley, CEDG decided to enter the competition, but it wasn't until about four weeks before the due date that we got serious about putting our submission together.

Because we did not have much time, it was essential to find quick and clever ways to produce two high quality presentation boards. The first board needed to identify assumptions, policies, issues, and strategies. The second was to show how to implement these ideas in the context of California's Central Valley.

We developed a team adept at using digital tools. Erik Peterson, an architect drew upon the masters thesis he and fellow graduate landscape architecture students did on integrating pedestrian-scaled communities into a natural context. We were able to use digital illustrations from this housing and put it in the context of the California Central Valley with the imagery that Jens Lerback helped produce. Daniel Cressy, who had worked on the Vietnam Orphanage Living Energy Systems, modified the systems diagrams and developed a profile to show how living energy systems could work in the context of a hillside development in the Central Valley. Erik von Wodtke, who learned Adobe Illustrator and Photoshop working on special effects in the movie industry, used these digital tools to merge the image of the cluster housing with a background of the Central Valley and put together the presentation boards. I wrote the narrative incorporating many of the concepts that our office has been exploring and laid out the boards. We produced the first board, which contained mostly text, using Adobe Pagemaker. The second board, containing mostly graphics, we laid out using Adobe Illustrator. After the project was completed and submitted, Erik von Wodtke transferred the files using Adobe Acrobat to create the PDF file that is included on the CD accompanying this book.

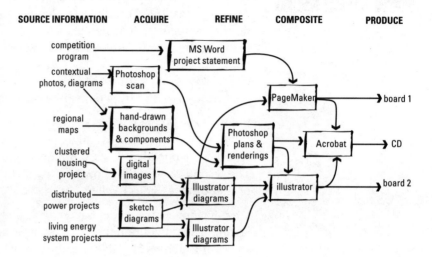

California's Central Valley
Competition information flow

The jury selected proposals for developing on flat land, providing infill that would extend the existing infrastructure. While these projects incorporated many of the admirable qualities of new urbanism, most did not address the issues of sustaining agricultural land, conserving water, and reducing greenhouse gases that contribute to destabilization of the global climate. We need to find ways to develop new human-scaled communities within natural boundaries of riparian corridors and flood plains. We also need to find ways to build strong regional economies that sustain and diversify agricultural production.

The proposal we submitted advocated regenerative designs that would preserve agricultural as well as natural services, provide clustered settlement on hillsides, and include distributed power production integrated into living energy systems. Clustering settlement on carefully selected hillsides could protect flat agricultural land in the valley. Hillside development has been practiced for centuries elsewhere in the world and has created beautiful settlements such as those found in Tuscany.

Changing from centralized power production to distributed power production could create more sustainable settlement. In the past 30 years, the development of the computer chip has stimulated a transition from central computing to distributed computing, creating the Internet and world wide web. Now, new industries developing photovoltaics, hydrogen fuel cells, and related technology could stimulate a transition to distributed electrical power production. Centralized power production uses polluting "dead energy systems" burning fossil fuels and nuclear power plants. Distributed power production can involve "living energy systems" using clean, renewable resources that do not produce greenhouse gasses. Renewable energy—which is already distributed—would be more sustainable, especially when fossil fuels become less available. Local communities would not have to be dependent upon importing fuel from transnational companies that risk the degradation of the environment by drilling for fossil fuel off the coast of California and elsewhere in the world.

Distributed power production also is more secure from natural disasters or economic disruptions that could interrupt the supply of fuel. A lesson to be learned from the recent war in Yugoslavia is the vulnerability of centralized water and electrical power infrastructures. California has one of the most vulnerable water and power infrastructures existing in the world today. There is an old saying that "those who throw stones should not live in glass houses." Settlement patterns in the United States resemble a proverbial glass house. By moving beyond new urbanism, we could create more secure patterns of settlement better integrated with nature.

Design professionals have a social responsibility to address issues facing society and provide the vision to deal with them. Digital media offer new avenues for design communication that could help bring about greater public awareness and mainstream implementation of ideas leading to a more sustainable future.

Summary
Guidelines for Visualizing Designs

1. **Keep design journals** to collect ideas and information. These should include sketches done on the backs of envelopes and napkins as well as in journals or diaries with photographs. Compile insights in electronic journals with digital notes, images, drawings, video clips, and even sound samples of key conversations to transfer them digitally.

2. **Develop models** to analyze, synthesize, and evaluate alternatives. At first these may be abstract concept diagrams of spatial relationships and processes, as well as images of the phenomena you are working with. Eventually models can become more representational with three-dimensional sketches and digital models, plans and sections developed over accurate base maps, renderings showing the image, and digital implementation documents. Use spreadsheet models to evaluate costs and revenues, as well as energy and water budgets to access the availability of key resources.

3. **Focus on the big ideas** and refine the content. Prioritize ideas and determine where to focus effort. Select the most appropriate tools and models to work with and refine the design process.

4. **Build teams** with the capabilities to deal with projects. Involve clients and consultants. Allocate responsibility.

5. **Identify appropriate audiences** (or markets) and think through how to tell them the story and share the vision. Find out the key parties involved in implementing the project, and understand what is meaningful to them.

6. **Network** with consultants, clients, and audiences. Remember communication works two ways—that which we can share, as well as that which we can glean from others through collaboration. Create shared value by building design equity, engaging stakeholders' intellectual and financial capital.

7. **Determine the most appropriate design communication tools** for the groups you are working with. This may be simple conversation in meetings or phone calls, online collaboration, formal presentations, or publications.

8. **Diagram information flow** to determine the best way to transfer information. Use models, transforming them digitally into appropriate presentations.

9. **Get feedback and refine ideas.** Remember, design is an iterative process and digital tools enable designers to work interactively.

10. **Develop preliminary designs into implementation documents.** Use digital tools to transfer information from schematic studies, through design development, to construction documents.

11. **Transfer what you've learned to the next project.** This involves not only content—the knowledge gained about the topic—but also the capabilities developed in using digital tools effectively.

Activities

Share the Vision of Your Design

Design professionals can enable others to share their vision by creating images for people to experience. They also can help develop and give expression to visions others may have. Images and other models offer a clearer picture of what a design will be—before it is built. This is crucial to gaining feedback and funding. Design professionals are unique in having this capability. We need to develop visualization and communication capabilities and use them responsibly. The challenge is to creatively communicate visions of designs that can benefit society.

Select issues you feel passionate about and follow through. This may be something of personal interest to you. Or it may be a local issue that affects your community—maybe even a global issue you can act upon locally.

A digital design journal enables you to work with online information, scan sketches and photos, or capture video and audio gathered in the field. Compile base information in digital formats so you can use it to accurately develop designs. Work smart, selecting the appropriate models for exploring the vision, and media for sharing it with others. Focus your creative energy on design content and communication that will be most effective. Network and build teams to share the vision. Use information technology to collaborate with others and engage their expertise. Interactively involve stakeholders who can help implement the ideas. Provide public service presentations and publish ideas—online as well as on paper. Enter design competitions. Cultivate clients who can make a difference. Use digital tools effectively to transform preliminary visions into real projects. Design with digital tools and communicate using new media, refining your expertise as you move onto new projects.

Digital tools and new media provide new ways to reach and interact with appropriate audiences. We can do well by doing good. What we create can have value. Everyone can benefit.

8

Multimedia: Reaching a Broad Audience

CDs and Digital Video

Design professionals have a need to communicate with many people. Often this involves pulling together the wide range of information and integrating it into a presentation that the public will understand and appreciate. CDs enable the audience to view multimedia interactively. With video the audience also can experience dynamic processes—such as changes that occur in landscapes or in human endeavors. Digital video can integrate a wide range of information and offers more opportunities to present the dynamics of change in ways that can be widely broadcast. Multimedia—CDs and digital video—have the potential to produce both excitement and understanding, thus providing memorable experiences.

The challenge is to do it well. James Sipes, a landscape architect with Jones and Jones (formerly an associate professor at the University of Oklahoma) has been able to use multimedia very effectively. He was the primary director of the CD produced for the Chattahoochee Riverway, a Landscape Architecture Foundation Demonstration Project that won a national landscape architecture merit award. He also has produced a video showing the impact of ice age floods on the Pacific Northwest. This video has been turned into a 30-minute TV special for the Oregon Public Broadcasting System.

Ice Age Floods
(see rest of sequence on p.144)

Jim Sipes especially likes to work with digital video, which—as he points out—is now affordable. A digital video camera and a computer with FireWire (a new standard for boards to input and output of digital video) for digital editing is financially within reach of many design professionals. The key is demonstrating the competency to make these digital tools productive. The public is used to seeing published CDs, broadcast-quality TV, and videos of movie productions. What we do as design professionals needs to meet those standards if it is to be taken seriously. Effective design communication balances content and information technology in a seamless way, without calling attention to communication techniques.

Just as design professionals have learned to work with hand-drawing techniques, ranging from quick sketches to finish renderings (or even with a camera, producing snapshots as well as fine photographs) we also can learn to work with video. It is relatively easy to use video to take field notes when gathering information for design. With some more expertise we also can use digital video for design communication. There is a growing genre of sketch video techniques—"shot from the hip"— as we see in independent films produced on low budgets. Of course there are more expensive polished productions for broadcast TV and film, but they are not necessarily better communication. As design professionals become more familiar with video, we will invent new ways of using this medium for design communication as well as for artistic expression.

Chattahoochee CD

The Chattahoochee River CD brought together the work of many landscape architectural groups. These included the Berkshire Design Group, Inc., Design Workshop, Terrence J. DeWan & Associates, Lee McMasters & Associates, Sipes/Leigh/Moore, Scott Banker & Associates, The Georgia Chapter ASLA Community Assistance Team, and the University of Oklahoma. The goal of the project was to focus attention on what could be done to restore the Chattahoochee River. The challenge was to integrate the work of these design teams into a coherent presentation that would be accessible to a wide audience in an interactive way.

The CD has a menu bar for navigation, enabling users to select different projects along the river, gaining quick and easy access to each section and their subtopics. Clicking on a project, the viewer will see images of the site and be able to access a design overview and concept plans. Music integrated into the CD provides a feeling for the place. The CD accompanying this book includes a segment of the Chattahoochee River project.

Chattahoochee River Project menus

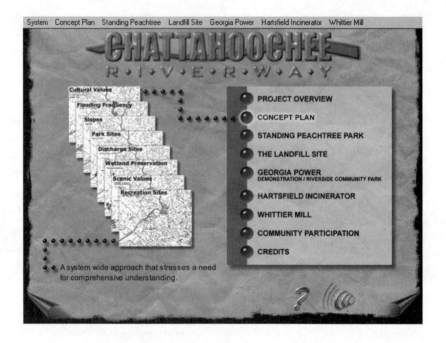

The menu served as a common interface pulling together the work of many design professionals addressing a variety of projects. This digital media links a narrative, an assortment of drawings, images, and even sound into a cohesive presentation.

Ice Age Flood Video

Video enables design professionals to take a large complex project— one involving all kinds of people—and integrate diverse information—showing everything from newspaper clippings and still photos to full-motion scenes from the site. We can incorporate images and animation, providing a vision of what a design could be, adding commentary, music, and other sounds. Multimedia is an effective way of engaging people and immersing them in what design professionals do.

To produce quality video, Jim Sipes points out that design professionals need the following:

• Quality input

• The capacity to manipulate images

• Quality output

And, of course, good design content and the capability to tell the story in ways that will engage the audience.

To capture live video, Jim Sipes uses a 3-CCD chip digital video camera. Although High 8 and Super VHS analog video cameras produce good-quality images, that quality is quickly lost once the images become edited and copied or when converted to a digital format. The advantage of digital video is that the images can be manipulated and reproduced without degradation. The cost differential between digital and analog video is rapidly disappearing. It seems that it will only be a matter of time before digital video will dominate and become well integrated with both computers and television. Exciting new products are available from Sony, Canon, and other manufacturers. Digital multimedia formats are emerging. MPEG compresses video and audio files that can be played using Apple Computer Corporation's QuickTime or Microsoft Window's Media Player.

polymorphic tweening *A change in morphology (or form) of an object, showing all the different shapes in between.*

To show full-motion images of what you can't see any other way, Jim Sipes uses digital animation. Although animation can be done tweening two-dimensional images (transitioning between key frames), Jim prefers to use three-dimensional digital models with which he can create a walk-through or fly-through. (3D animation is actually much easier than 2D, especially for beginning animators.) He also can dynamically change the objects and add turbulence and chaos to make the images look more realistic. For example, he animated the salmon run along Pacific Northwest Coast for a project dealing with habitat preservation. He also animated ice age floods showing the massive flows of water released across northwest America when ice dams broke and reshaped the land. This animated video provides a memorable understanding of the forces forming this landscape. Jim Sipes' 13-minute animated video is available through the National Park Service (NPS). Seeing this video, the Oregon Public Broadcasting System asked Sipes to work with them on a 30-minute TV special about developing a resources master plan protecting this landscape. Parts of this video can be seen on the NPS web site www.nps.gov/iceagefloods

Multimedia Production Process

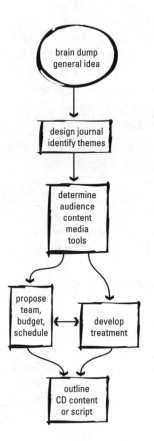

Jim Sipes likes to began his projects with what he calls a "brain dump" to generate ideas. He typically gathers his first insights on paper using key words or simple pencil doodles. From that, he is able to identify issues, define problems, and scope out what needs to be done. He will usually develop a journal in a word processor to start the creative juices flowing and save the ideas so that he can let them ferment. From this, he gets a sense of what needs to be accomplished and interactively begins to look at the best ways to get it done. The key is to generate creative ideas that become the basis for selecting the content, tools, and media to reach the audience. Is it better to produce an interactive CD for this audience? Or, is it better to produce a video for projection or broadcasting?

At this point in the process, reality must set in. It is necessary to look at the time, budget, and human resources that are available. One needs to develop a strategy to deliver the presentation within these constraints. This strategy is often expressed as a one-page description or "treatment" for the production as well as a working outline of the content or script. The value of a multimedia production needs to justify the expense. It usually is necessary to find sources of revenue to add to what is typically budgeted for design communication. These may be funds for public information, promotion, permit processing, and marketing. It makes sense to proceed only if the necessary time, funding, and team is committed to producing a quality product. Multimedia projects can range from simple to more elaborate ones incorporating advanced technology. Both can be effective. Design intent and budget are important in determining how to proceed.

Preproduction

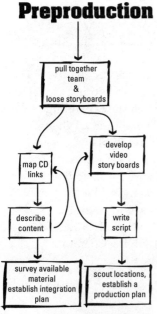

Production

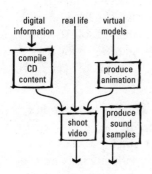

Like with a construction project, once there is funding we can proceed with a preliminary design. Preproduction begins with a loose storyboard that establishes a basic framework to provide focus and save time, but Sipes prefers to keep it flexible so that he can respond to opportunities that may arise during production. He may begin a loose storyboard with 3 x 3 inch yellow sticky notes that he can move around on paper, or work with a computer graphic program, cutting and pasting images of key frames to explore the story continuity and transitions. When producing a CD, it is necessary to explore links in the information environment that the user will navigate. When producing a video it is necessary to develop the script and refine the storyboard. Using PhotoShop enables Jim Sipes to scan images of key frames to develop the storyboard. Using a word processor for the script and developing the storyboard digitally enables design professionals to share this information online with others on the team, as well as with the financial backers of the multimedia production. It also provides the basis for the visual, financial, and organizational planning necessary to successfully produce the project.

Production involves compiling the content for the CD, producing the animation models, or shooting video footage.

The content for CDs can come from many sources, but must be compiled using a format recognizable to the browser. Microsoft Explorer and Netscape can access HTML files placed on CDs. Adobe Acrobat enables a user to navigate PDF files that are more compact and have a more consistent layout. Some CDs use proprietary interfaces that programmers can develop to enable the user to access the content in many different ways. Sipes points out that producing a good CD involves more production effort than most people realize.

Animation involves producing a sequence of images that enables the viewer to experience movement. This may be based on a sequence of 2D images tweening (transitioning between) key frames. Or, it may involve building a 3D digital model through which observers can move or view as it transforms. Digital images and 3D models can lend themselves to animation. It's a matter of using the synthesis models that come out of the design process and taking them further to produce animation for design communication. While this capability is emerging within traditional design professions, as Jim Sipes demonstrates, service bureaus are emerging with a new cadre of digital artists who can render, animate, and even add digital special effects to models designers produce. (Some of these people were educated in architecture and other traditional design disciplines, and have used their digital modeling skills to transition into the entertainment and computer graphic industries during the last recession.)

Digital video usually involves using a camera to shoot footage, but we can also capture digital video from computers. This may consist of stills or full-motion video. With digital video cameras, shooting non-linearly (out of sequence) is not a problem, because we can quickly access scenes and easily edit them later with no loss in quality. This can simplify shooting footage. With analog video, shooting the scenes linearly (in sequence) could minimize edits, but more planning has to go into the camera work to maintain pictorial continuity. When shooting any video, however, it's very important to plan scenes so they relate to the story that is being told and maintain the subtle continuity of light and movement necessary for images to flow together in convincing sequences. In some instances, the story may unfold in front of the camera, providing opportunities for the production to evolve. Sipes acknowledges that there is a delicate balance between planning a production and staying open to possibilities as they occur in front of the camera.

Postproduction

Postproduction involves putting the pieces together and previewing them before they are output.

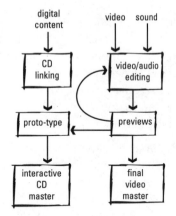

When working with CDs, the pieces are put together according to the map of the CD's information environment which needs to be designed in a way that enables the viewer to interactively navigate the content. Web authoring tools and programs like Adobe Distiller enable designers to compile content and establish links. The links should be tested to make sure that they all work and are reasonable to follow. Sipes points that this last part of the project—getting all the final bugs out so the CD will run on multiple platforms—can take even more time than the first part of the project.

When working with digital video, the production is put together according to the script and storyboard to provide seamless transitions through the various scenes that tell the story. Jim Sipes uses Adobe Premier for editing. Although digital editing software may provide fancy transitions, it is often best to keep the production simple. Sipes likes to make technology transparent and focus on telling the story visually. Multimedia programs also provide the digital tools for integrating sound tracks.

Postproduction of digital video involves inputting video from a camera and storing it on a hard drive, editing segments with editing software (such as Premiere), rendering the final sequence as a compressed animation file, and copying the animation file back to a digital master tape.

When editing video, Jim Sipes will render out previews at one quarter size (320 by 200 pixels) to get the timing. This keeps the files small so that he can work more quickly. He may do six previews of short transitions and then preview the whole production maybe four times to get an idea of how it progresses. Using FireWire, a computer with a fast 10-gigabyte hard drive can run about 15 minutes of digital video if it is reduced and compressed 75 percent. It is important to archive the digital production files on CD along with the video editor's definition file that puts them all together in a timed sequence.

Currently, Jim Sipes will produce the final video at a resolution of 720 by 480 pixels with 24-bit color, using QuickTime for CD production or AVI for digital video. This resolution is better than current broadcast-quality television. Sipes typically reduces the format to half that size for CD production to keep the files smaller. For digital video, he will make a master on DV tape that can store a production of up to 70 minutes. This master can be used to make analog video tapes using the typical VHS format. Storing finished digital video in an API format involves a lot of disk space. As digital video disk (DVD) recorders and players become more available, DVD will provide a format with sufficient space for storing and viewing digital videos at even higher resolutions. Then we can present drawings with the finer line work that we like to use in the design professions.

Digital multimedia is a useful vehicle for design communication. Effectively using digital tools in this way could also enhance the visibility of the design professions. Broader bandwidth communication will enable us to transmit video online. We are just beginning to see the potential. Jim Sipes and I are already looking forward to the time when we can use holograms to project images of three-dimensional models displaying designs in the air to an audience surrounding these digital information objects.

Summary
Guidelines for Using Multimedia

Development "What are we going to do"

1. Collect ideas and insights in a design journal. Identify themes, issues, problems, and opportunities, and scope out what needs to be done.

2. Develop a treatment describing your strategy for delivering the content to your the audience.

3. Propose a team to obtain funding for producing the project.

4. Outline content of the CD or video script.

Preproduction: "How are we going to do it."

5. Explore strategies creating a loose storyboard, and/or diagramming links for producing an interactive CD.

6. Develop a storyboard to plan camera shoots and develop animation.

7. Describe the content for interactive CD, or write a script for video, telling the story to your audience.

8. Plan how the team will carry out the production.

Production: "Doing it."

9. Compile content for the CD and transfer it into hypertext markup language, Adobe Acrobat, QuickTime VR, or other useful file formats.

10. Produce animation sequences enabling the viewer to experience movement.

11. Shoot video footage from real life, or from computer images and animation, and produce sound samples.

Postproduction: "Putting it all together."

12. Link the CD content in ways that users can navigate interactively using Adobe Acrobat, MS Explorer, Netscape, or other browsers.

13. Composite the video, editing footage with transitions that provide story continuity. Add sound tracks.

Activities

Use Multimedia for Your Project

Who is the audience you would like your designs to reach? Can you build awareness of issues and offer visions for dealing with them? Can you effectively communicate the design during each stage of a project's approval process? Can you help engage stakeholders to implement the plan? Can you provide services that will help market the design project, or market the design services you provide?

Select a project where you could use multimedia to reach a broader audience. Draw upon digital models you've already developed for the design. Look for funds to produce a CD or digital video. Build a team that can deliver a quality production and follow through using the guidelines provided in this chapter.

Keep in mind that digital tools for multimedia will rapidly evolve as digital cameras and recorders, computer technology, and television merge. Digital video disks (DVD) and high-definition television (HDTV) will improve the available storage capacity and resolution, making digital technology an effective vehicle for design communication. This, coupled with broadband communications through digital service lines (DSL), cable modems, or satellite transmissions, will enable those who develop multimedia to deliver their content quickly and easily. While the telephone, FAX, and e-mail are information technologies design professionals now use for collaboration, multimedia may become another important collaborative tool as well as a means for reaching a broad audience.

9

Applying Special Effects

Digital artists create special effects that can have tremendous emotional impact. The movie industry needs to tell stories in ever more compelling ways. Advertisements must make their point in 30 seconds or less. There is tremendous potential for using special effects for design communication, conveying realistic visions and images of places and products that may not yet exist. Film, digital video, and computer-generated imagery (CGI) are collaborative commercial arts. This collaboration involves producers working with directors who have the capabilities to tell stories visually, by working with artists who can create, edit, and composite 2D images, and create 3D animation.

Gray Matter FX

Responding to a growing demand for digital special effects, Gray Marshall, a visual effects supervisor, and Margot McKay, an executive producer, established Gray Matter FX in Venice, California. They provide digital compositing to commercial and feature film clients. They seek a creative relationship between the client and the compositor, using a variety of digital tools. Talking with Gray Marshall provides insight on how digital artists are now able to work, using the new tools that are emerging.

Gray Marshal

Before establishing GRAY MATTER FX, Gray Marshall worked as a compositor and visual effects supervisor at Digital Domain for 4 years. Digital Domain is the studio James Cameron founded that did the movie "Titanic." At Digital Domain, Gray Marshall helped set up production procedures and beta testing in-house compositing. He was the visual effects supervisor for "True Lies" starring Arnold Schwarzenegger. This feature film was noteworthy in that it used spectacular special effects to tell the story in ways that made it difficult to determine what was real and what was not.

Gray Marshall describes his background in terms of acquiring both "how to" and "why to." He began developing the "how to" when he was introduced to computer programming as a teenager in 1976. He studied radio and TV production at George Washington University. Driven by a fascination for computer graphics, he went on to Pratt to study computer science, where he got involved in fractal simulation. Learning to program enabled him to "weave his own yarn and not be limited by what is on the shelf," in other words, to become a master of the medium and not just a user. Seeing "Luxo Junior," a computer-generated film, at SIGGRAPH 1986 (the Special Interest Groups in

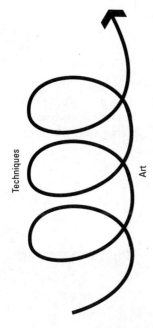

Techniques

Art

Upgrade cycle

Special Effects Studio

Graphics computer show) helped him realize the artistic potential of the medium. The "why to" is simply a love of cinematography, evolving from a minor in art history at George Washington and continuing through a masters program at USC Film school where he learned to tell stories visually. He had an internship at Industrial Light Magic where he began as 3D animator before working for James Cameron at Digital Domain.

Special effects have become a collaborative art form that involves many people. This digital art includes cleaning up the base model, texture painting, manipulating lighting, color, and viewpoint, as well as animation, miniatures, motion control photography, props, and pyrotechnics. Gray Marshall observes that few people accomplish all the skill sets, or should they. It is a fallacy to think that because the program can, the person who runs the program can. And just because people can, it doesn't mean they should. A good production takes developing a team of digital artists—people with techno-savvy and artistic talent. Chasing the upgrade cycle makes it difficult to balance technique and art. It is important to find time for both. Successful productions require collaboration that, of course, also involves the producer and director, as well as digital artists working through an iterative process.

Gray Matter FX uses Discreet Logic Software (Flame, a compositing tool, and Smoke, an editing tool) on Silicon Graphics workstations he leases for production. Leasing equipment makes it easier to continuously upgrade evolving digital tools. Their work involves storing massive digital video files on large hard drives. They back up on tape. The trend is "cheaper, simpler." They are now starting to use Shake, a new compositing software package, and Elastic Reality software that can run on a Windows NT workstation. In addition, they have the typical business computers to run the office and correspond online.

Video currently takes 30 frames per second, although there is a movement to change this to 24 frames per second—synchronizing it with film and reducing files size. Because each frame for a commercial-quality production uses approximately a megabyte, this level of digital video may result in about 30 megabytes per second. Using compression and lower-resolution drafts can reduce file size, but the files are still huge. Film frames are as much as 10x larger than video frames. Gray Matter FX almost never uses compression, but sometimes does work out ideas at a proxy (reduced) resolution to increase speed of iterations. Film frames at 10 megabytes each and running at 24 frames per second creates a massive storage problem. It takes careful file management to work effectively and keep track of the work as it develops through each iteration of the process.

The Production Process

Preproduction involves developing the treatment, script, and storyboards or animated preview shots to assign timing. Production involves filming the scenes or shooting with a digital video camera. This may be background plate (BG plate) or foreground (FG)—live action (often shot in front of a green-screen). Production may also involve developing animation and recording music and other samples for the soundtrack. Postproduction involves cleaning up and compositing images carrying out the special effects and mixing the soundtrack.

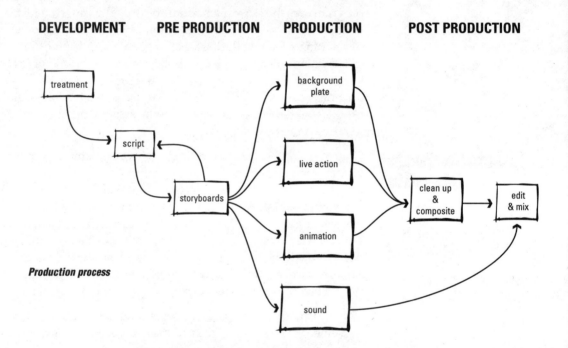

Production process

Special effects need to be carefully planned in preproduction to determine the best approaches and techniques for each project. Production requires the careful gathering of the raw material. The real magic, however, happens in postproduction where digital artists work in an iterative process that leads to the final product. This can involve collaboration with design professionals who may provide content for the digital artists to use for creating special effects to refine a presentation. A recent book by Ron Brinkmann, entitled *The Art and Science of Digital Compositing*, further describes how digital artists work. A few examples from this book are included on the accompanying CD.

The Models

Each frame involves elements on layers that can be edited and composited digitally. Similarly, sound sources have samples kept in tracks for editing and mixing. The BG plate provides the context for the scene. The action may involve the actors recorded in front of a green-screen, or animation produced virtually as 2D or 3D objects using digital tools. The "matte" combines the plate and action in a composite that can be synchronized with the sound track.

In "True Lies" James Cameron filmed a Harrier aircraft with actors hung on cables from a crane in front of an office building. In postproduction, special effects artists removed the cable, added visual heat effects for the exhaust. The sound track added the roar of the engine, music, and dialogue.

The Pipeline

Gray Marshall refers to the information and workflow as the "pipeline." He recognizes how important it is to have a clear picture of what you are doing, especially when working collaboratively on something as expensive as a feature film or commercial production. He observes that for the best integration, the visual effects supervisor should be brought in early to help plan the shots needing the effect. The visual effects supervisor needs to clearly communicate how to make things happen.

The pipeline may begin with a shot that becomes the backplate. This can be handed off to different disciplines who may add the live action shots—using green-screens, 2D or 3D animation—and sound.

Pipeline diagram

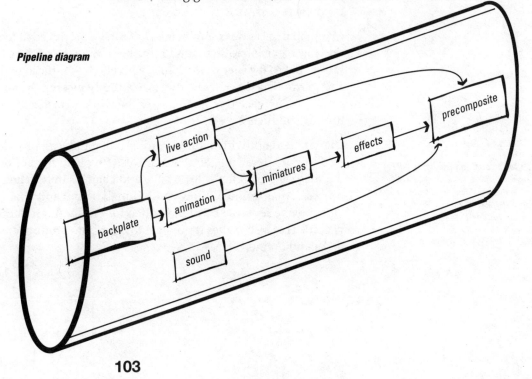

103

Additional filmed elements, such as live action shots, may or may not be on green-screen. These elements include BG plates, FG green-screen, smoke, debris, crowds, miniature vehicles, and landscapes. There also may be 2D generated elements such as rotoscoped mattes, effects animation, and precomposites. There may be 3D computer-generated elements as well to provide whatever is required for digital special effects.

rotoscope *Tracing and cutting out an image to composite for special effects.*

background plate (BG plate) *The back plate for live action (often shot in front of a green-screen).*

All of this gets composited, bringing the elements together. The tape is then sent out for approvals. The director and producer need to approve the scenes, motion, and sound before the animation artists can finish production. Initial reviews may use lower-resolution proxy tapes of digital works in progress. Later iterations, involving more detail, may be on film. There may be two or three "film outs" (where digital tapes are transferred to film) before the final.

In doing work on the "X-Files" Gray Marshall worked with live-action shot on a building roof in Los Angeles, rotoscoped and integrated it into a BG plate of Houston, Texas. The enhanced action in the result looks like it was shot in Houston, demonstrating that it is now possible to use new media to create new realities. Gray Matter FX does this everyday.

Interactive Iterations

Gray Marshall talks about more interactive iterations to do new things, and explore techniques and visions (from both the artist and the director). This takes an investment in time, but can pay off. It is a commitment necessary for continuous growth.

Marshall likes to do commercials that can be produced at lower resolution for television rather than for the big movie screen. The lower resolution makes it easier to work interactively in the studio with the client, and permits trying out new things. Carefully picking the appropriate resolution for each operation makes the work go faster, saving editing time and reducing costs.

high-definition television (HDTV) *The new standard for digital television that has higher resolution and a wider screen than conventional analog color television.*

Digital media provide more control. Analog film can produce higher resolution. Interactive iterations are also involved in handoffs from digital to analog. Film is expensive and changes in the chemical process. Analog video is low resolution. Film out and analog videotape may be replaced by digital video for HDTV (high-definition television). This also makes it possible to distribute productions on digital disks and online.

Value
Special effects production costs raise interesting questions regarding values. We may not be able to afford to do it just because it can be done. Value must justify cost. Initially, people thought the movie "Titanic" was going to sink financially. This production made extensive use of expensive special effects that helped it succeed in reaching a large audience and have a memorable impact. As we continue to see more special effects used to tell the story, we will see more extensive use of digital tools.

Ethical values also need to be addressed. While "true lies" may be appropriate in movies where people expect fantasy, the manipulation possible with new media may not be acceptable in political messages, advertising, and design communication. It remains to be seen how society will deal with the distinction between reality and virtual reality.

While deception is as old as communication, new media make it possible to create whole new realities. Multimedia, however, are very revealing. The audience gets the message in many ways, and if these cues do not combine credibly, the audience may feel deceived.

Summary Guidelines for Using Special Effects

Development

1. **Develop a clear picture of what you want to do.** This should include a strong story you want to tell, a treatment for delivering this content to your audience, and funding for producing and getting it to market.

2. **Determine what special effects would be useful.** Talk with a visual effects supervisor who can help visualize how the special effects could be used and developed.

Preproduction

3. **Develop a script and storyboard.** Figure out how to use special effects seamlessly to tell your story.

4. **Pull together a team** with the technical savvy and the artistic talent to produce the special effects. A special effects studio, with a visual effects supervisor, talent, and tools can help carry out the project.

5. **Plan the production process.** Careful planning makes production more cost-effective, and makes it possible to carry out quality special effects in postproduction.

Production

6. **Get the right digital information into the pipeline.** Produce the digital models or record the live action needed to tell the story. Make sure you have good raw material for your special effects.

7. **Determine the appropriate resolution** and make sure the digital information transfers effectively to the final product given the equipment you are using.

Postproduction

8. **Clean up and composite the digital information** to create the special effects.

9. **Work interactively through iterations to get it right.** Digital tools make it possible to interactively work through many reiterations collaborating with the digital artists. The trick is to be able to do this quickly and effectively.

Marketing

10. **Integrate the special effects into the production and recover the value** by delivering it to market. Typically, special effects need to reach a large audience or contain a valuable message to justify the cost.

Activities

Use Special Effects for Your Project

Digital artists create special effects to help tell a story. The challenge is to use electronic media in ways that become transparent. Tell a story—relating the experience—without bringing attention to the effects.

Select a project that needs special effects to tell a story that gets the message across, or communicates a design. Think through which special effects would be useful. Will this involve a single image, a short video clip or commercial, an infomercial or documentary? Or, could it even be part of a feature film? Determine how to fund this production and get it to market.

Use the guidelines in this chapter to help build a team that can deliver the quality of production you need. The real trick is balancing the costs and the value of the production. Delivering it on time is another challenge, since producing special effects can be very time consuming.

Connect with a distribution network to get the message to the audience so that the value justifies the costs. Today, more channels are emerging for distribution. They include:

Websites

Cable or broadcast TV

Video distribution

Film distribution

Design professionals can work with digital artists, writers, and directors to provide content. Producers and publishers can help pull this content together to get it to market. The World Wide Web makes it possible for design professionals, digital artists, and others to produce content independently and get it to appropriate audiences.

10

Applying 3D Modeling

NBBJ Sports and Entertainment/ Los Angeles

Using 3D digital modeling, design professionals can demonstrate to clients their capability to create appropriate and interesting spatial experiences in ways that are readily understandable. These 3D digital tools enhance our ability to design complex forms. Digital models also can help clients communicate with a wide-ranging constituency to market facilities before they are built and engage stakeholders for funding and public support.

NBBJ, the fifth largest architectural firm in the world, has successfully implemented digital 3D models into the process of designing sports and entertainment facilities. Michael Hallmark AIA, Principal of NBBJ Sports and Entertainment, points out that, on a building with relatively few straight lines, digital 3D models are essential for understanding design ideas. They provide tools to study volumes and form and as well as refine detail. The interior concourses and club suites (areas highly scrutinized by the client) can be extensively modeled with colors and textures scanned from materials intended for the space. In addition to helping select materials, the digital 3D model can aid dialogue with fabricators who need a full picture to build the complex geometry. Paul Q. Davis, an architect who is a senior project designer for NBBJ, observes that some of these forms would be extremely difficult to design and build without digital 3D models.

On the cover of this book is an image of the Paul Brown Stadium, designed by NBBJ architects, for the Cincinnati Bengals football team. The image depicts one of four distinct circulation ramp towers that are situated at quadrants of the stadium. The image represents a snapshot in time of an ongoing digital 3D model that was used primarily by the design team for testing the validity of various ideas. Doubling as design communication, this particular image was also instrumental in conveying the intent to the client. Also pictured are the stadium's VIP entry and a glass-enclosed stair tower along with precast concrete and perforated stainless steel skins that veil movements along the stadium's pedestrian ramps. Images like this were later set against a backdrop of the city to evaluate the subtlety of building colors against the skyline of Cincinnati.

Paul Davis related to me how they went about doing some of the digital 3D modeling for the Paul Brown Stadium. Davis comments, "We truly use it to investigate architecture." Sight lines optimizing the view from each seat in the stadium help to develop the complex form of the bowl and intricate geometry involved in the roof. "It is like boat hull design."… "Every bay changes."… "Computer 3D models provide a way to understand the design." These architects use digital 3D models as part of preschematic design proposal to win the job, then for the preliminary and design development, as well as construction documents to help work out fabrication.

Digital 3D model work flow

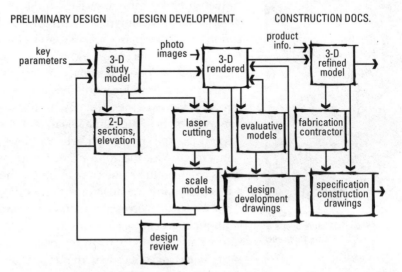

Preliminary design modeling begins by working with key parameters, such as sight lines, to generate raker beams, boomerang columns, and other components to create the roof form. With this, the architects can produce a digital 3D study model to optimize the design and integrate it with a digital model of the site. From the digital 3D study model they can quickly generate accurate plan, section, and elevation drawings to examine the design in more detail. This information, along with the 3D model, is shared with the structural engineers who do the finite analysis to determine the size of structural members. The digital 3D model is reworked again and 2D CAD drawings are generated for a physical scale model. Digital drawings, transmitted by modem to a model maker, are fed into numeric-controlled laser cutters to accurately create the Plexiglas parts for physical models at any scale the architects wish to build. Parts, exact to the computer model, can arrive the next day and assembly of a physical model begins. Physical models help visualize how the pieces go together and are a useful display for clients and their constituency. The feedback obtained from the client and other parties involved in the project can be incorporated into the digital 3D model that becomes the core of the preliminary design. Plans and sections provide a reference for quick measurements, horizontally and vertically along gridlines.

109

Paul Brown Stadium
(NBBJ Architects Sports &
Entertainment)

In the design development stage, the preliminary digital 3D model can help consultants even more. Structural engineers use this along with digital plans and sections to extract information for further analysis to refine the size of structural components and begin detailing connections. Mechanical engineers can use the 3D digital model to avoid interference when routing a myriad of pipes and ducts for plumbing and HVAC systems through the structure. Architects and engineers—transmitting key drawings back and forth attached to e-mail—can collaborate as the design process progresses. Sometimes quick communication through e-mail can solve issues between multiple parties while the paper documents are held up at the printers, are in transit, or are rolled up on a desk that is inaccessible. Using the 3D digital model, the architects can examine sight lines from seats where railings may interfere with views of the arena. By simulating views from selected vantage points in the digital 3D model, architects can verify which situations may be critical and help convince the client to spend the additional monies for the appropriate solutions. The digital model also helps create value by helping optimize the location of concession stand advertising, and the positioning of scoreboards for optimum viewing. Architects can even explore solar orientation by simulating rays of sunlight at crucial seasons and times of the day. They can visualize colors and textures by mapping materials on the three-dimensional form. Paul Davis likes to scan portions of photos of actual materials to simulate colors and textures. Using these techniques, he simulated the translucency of roofing material to convey the different feeling between a translucent fabric roof and an opaque metallic roof for the Paul Brown Stadium. All of this is helpful in selecting colors and materials and specifying components.

In the construction document phase, the architects are able to explore—in three dimensions—the components of the building skin and examine how they connect. Using the design development 3D digital model—rebuilt with more accuracy and a greater concern for detail—sections are taken from the computer model to derive working points for the structural engineer to work from. They also help the contractor understand the complexity of the project. On the Paul Brown Stadium, the roof model (about 1 megabyte of data) was attached to e-mail and sent back and forth between NBBJ and BIRDAIR (fabric roof specialists/contractor) three or four times In this way the architects were able to work with fabricators figuring out how to roof the curved forms. Often, Davis has found himself using the 3D digital model to show fabricators how the pieces must fit together and provide them with accurate dimensions in a 3D frame of reference. What excites the client may scare the contractor. There is a tendency for contractors to submit extremely high bids on complex facilities to protect themselves,

Paul Brown Stadium Roof
(NBBJ Architects Sports &
Entertainment)

and then they endeavor to simplify the structure in ways that may compromise the design. Davis observes that considerable money can be saved once contractors understand the complex form and have confidence in how they can build it. Digital 3D models help to drastically reduce their cost estimates.

Cyber Meisters

A lot has changed since the 1980s when Paul Davis went to architectural school and professors critiqued his 3D digital designs—commenting mostly on what the computer did (or often could not do at the time)—almost as if the designer didn't matter. Now, offices and schools throughout the country are developing a better understanding of how to use digital tools for three-dimensional design. The tools themselves are becoming much better and more affordable. Today, Davis talks enthusiastically about "cyber meisters"— his colleagues who have mastered the ability to design in cyberspace. These include Joey Meyers who designs with a SGI workstation at NBBJ's main office in Seattle. Jonathan Ward uses SGI and Microstaion Triforma at NBBJ's office in Oslo, Norway. In NBBJ's Los Angeles office, Rania Alomar and her team were responsible for the digital design work done for the new Staples Center. A growing network of people designing with digital tools is emerging in offices, online, and on the World Wide Web. Designers can help each other and advance the state of the art of using digital tools throughout the design professions.

Staples Center

Staples Center, opening in Los Angeles, will compete effectively with New York City's Madison Square Garden, attracting events such as the Grammy Awards and major entertainment. Staples Center also is a sports arena for the Los Angeles Lakers, Clippers, and Kings. This entertainment and sports facility, with its record-breaking sponsorships, is becoming a new benchmark from which future projects of this type will be measured. Its three levels of luxury suites represent the latest development in the arena planning and design.

Staples Center (NBBJ Architects Sports & Entertainment)

111

Michael Hallmark recognized the need for doing this project in downtown Los Angeles to develop NBBJ's Sports and Entertainment practice. To get this job, NBBJ ventured thousands of dollars in architectural services to develop a presentation that would distinguish their firm from the competition. Building upon what they had learned with other projects, such as the Paul Brown Stadium, NBBJ used 3D digital tools to quickly build a model of the downtown, and a model of the project site, along with a preliminary design study for the new center. They captured images of Los Angles along with those of sports and entertainment events that could help convey the spatial experience one might have visiting this new center featuring their new concourse concept. They integrated all this into a video presentation with an animated fly-through of the fluid forms of the stadium integrated with the massing model of city set to a throbbing musical beat. In addition, using the digital models to quickly laser-cut the pieces, they produced a physical scale model that was almost 8 feet by 8 feet. This presentation showed both their innovative thinking as well as demonstrating how they design buildings and communicate. NBBJ's effective use of digital tools helped them get the design commission.

Reaching the Constituency

The developers were quick to recognize the promotional value of this approach. They had to reach a wide-ranging constituency including the local jurisdictions responsible for approvals, financial stakeholders, as well as the market that would lease the spectator suites and engage teams and events. The developers used the digital imagery that NBBJ produced in unprecedented ways to promote the project.

Three-dimensional models helped to gain approvals by providing the community with a clear picture of the project. Developing the design digitally, and using this soft prototype as a basis for cost estimating, helped to evaluate the costs with greater accuracy. The digital models also made it easier to work more interactively, optimizing the design as it progressed through the process.

Videos—presenting virtual three-dimensional models and images— help to market the suites, long before physical models could be built for promotion. The videos could convey the excitement of experiencing an event at the center in ways that would not be possible using static physical models. Video seemed particularly appropriate for marketing to the Los Angeles region— known for film and TV. The design team also explored ways to integrate advertising into the facility through a digital scoreboard and other devices, adding to the potential revenue sources.

Optimizing the Process

Paul Davis observes that even more could be done with the digital information. Using 3D digital models—already being created for these projects—interior architects and other specializations could further explore their aspects of the design. These models could become the focus for the collaboration and enable the team to more interactively refine the design throughout the process. Coordinating these efforts online could take time-lag out of communications. This would enable the team to get beyond some of the current cumbersome practices built upon "phone tag" and "paper trails" which can be all-consuming on projects of this scale, while still maintaining the necessary documentation of decisions. Ultimately, more 3D models and other digital information could be available online for consultants to download using FTP (file transfer protocols.) By collaborating online, everyone could immediately have access to the latest information when contributing their expertise. For this to happen, however, will require high-speed transmissions and collaborative teams capable of transferring digital drawings and 3D model files.

Digital 3D also provides tremendous potential for relating the design process to the construction process. As Davis observes, "Everything I did to explain the drawings (for the Paul Brown Stadium) came from the 3D model." If both the architects and the contractors had digital capabilities, the specifications and digital 3D models could become the primary document. Details and shop drawings based on this digital information could provide clarification. This might replace the volumes of construction drawings now produced on paper that are now the primary part of the construction document packing. Having contractors work from 3D digital models could offer a more accurate description of the project. Curved forms are hard to accurately describe with 2D drawings. Digital 3D models would enable the contractor to use digital tools to query the drawings to get accurate dimensions and take-offs. The contractor could also make digital as-built documents that could be valuable for facilities management.

NBBJ moved the Staples Center Project team to a trailer at the project site during the construction of the facility to stay in close contact with the construction process that was on a very fast timeline. Through telecommunication, this site office could stay in touch with the resources they needed in the NBBJ offices. The office also provided a website so anyone in the firm, as well as the clients, stakeholders, and even the public could visit the project online and see digital images of the progress.

Summary

Guidelines for Using Digital 3D Models

Preliminary: "What is the form?"

1. Select a project (or objects from a project) where a digital 3D model would be useful. Commit to developing a team with the digital tools—needed to design in 3D—to create models that will be useful.

2. If the project is site specific, develop a 3D model of the site as a base for integrating the built objects. Determine the parameters for generating 3D objects digitally.

3. Develop the digital model to explore form. Simulate sight lines, sun angles, and other key determinants. Interactively refine the form. Also use the model to do takeoffs for budgeting purposes and optimize the design.

4. Use the preliminary digital 3D model, along with sketches and images, to communicate the basic form and feeling of the design to clients, stakeholders, and review bodies. Where appropriate, incorporate their feedback into the digital design model.

Design Development: "What are the materials?"

5. Share the digital 3D model with consultants. Encourage them to use the three-dimensional base information and assemble their work in the 3D model to see how it fits.

6. Refine the model adding materials and examining connections between key components.

7. Use the developed digital 3D model to communicate the color and materials of the design with clients, stakeholders, and review bodies. Incorporate their feedback into the digital design model where appropriate.

8. Make 3D digital information available to the client for promotional purposes if appropriate.

Construction Documents: "How will it be built?"

9. Share the model with fabricators and contractors and work interactively to resolve the best way to build the project.

10. If the fabricators and contractors are willing and prepared, provide a 3D digital model along with specifications to serve as the construction documents. Use details and shop drawings for clarification where necessary.

Construction Administration: "What is built?"

11. Use the digital 3D models to record changes reflecting the as-built conditions. Make this digital information available in formats appropriate for facilities management as well as for future additions or renovations.

114

Activities

Design with Digital Models

Select a project to design with 3D digital models and use the guidelines to help you. Now it is possible to design with digital tools exploring form with 3D models. Some projects, involving compound curved forms, are most effectively designed that way. Enhance the value of this approach by using the 3D digital information in many ways throughout the design process.

A key challenge is to develop a team with the capabilities to work with digital 3D models. While 3D design communication may actually be easier for a wide range of people to understand, it presents challenges to those doing the design. Most value is derived if the design team, including consultants, can all work with the 3D digital information, and incorporate their work. It would also be extremely beneficial if those involved in the fabrication or construction of the designed objects could also work from the 3D digital model. Given that this is a fundamental paradigm shift in how the construction industry currently functions, it may take time to change. In manufacturing, these changes are already occurring, providing manufactures—that have this capability—advantages in both the speed and quality with which they are able to get products to market. These competitive advantages are likely to drive change in the construction industry as well—especially for complex buildings such as sports facilities.

11

CADD:
Communication
Aided Design
and Data

"The information revolution is proving to be more about communications than about computers. In one industry after another the integration of on-line communications into traditional business models is transforming the desktop computer from the locus of business intelligence to a node where humans interact with a vastly more intelligent network."
—Jerry Laiserin, writer for CADENCE Magazine and an editor of ACADIA—the journal of the Association for Computer-Aided Design in Architecture

Coordination and Control

Design professionals coordinate a considerable amount of information as part of the design and implementation process. Project-based digital information sites provide a vehicle for expediting information flow and smoothing workflow to improve the overall quality for everyone involved—design professionals, consultants, clients, stake holders, and even review bodies.

Often design professionals only optimize some facets of the design process—such as computer-aided drafting—by implementing more "tips and tricks" but following the same old procedures established for hand-drafting. Greater gains can be derived from broadly applying digital tools—throughout the entire design and implementation process—more effectively using equipment we may already have invested in. To do that requires rethinking the process we use for computer-aided design. Initially CAD stood for computer-aided drafting, but as it embraced more of the design process it became known as computer-aided design, or CADD—computer-aided design and drafting. Now, Jerry Laiserin—an architect and industry analyst—observes that we are seeing "a new kind of CAD—*communication-aided design.*" This involves using information technology to design with digital tools and share data throughout our entire enterprise. Perhaps CADD should now stand for *communication-aided design and data*, to more completely reflect this paradigm shift.

One of the keys to sharing digital information is coordination. Tony Palmisano—an architect who worked for Wolf, Lang, Christopher (WLC) an architectural firm in Rancho Cucamonga, California—took the initiative to organize a meeting for all their consultants to discuss ways for improving coordination. I was invited to participate in this coordination meeting because CEDG, consults for WLC. Palmisano was seeking better ways to share information between architects, consultants, and also more directly involve clients and owners.

To do this, he advocated setting up and coordinating project-based information sites to develop more of a "project point-of-view" as a focus for collaboration. This means moving beyond working independently and handing off (or express mailing) drawings or disks with digital files. It also involves more than attaching files to e-mail. Project-based sites provide a place to share and keep track of project information online. This requires setting up project-based FTP (file transfer

116

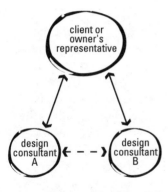

Owner-coordinated project

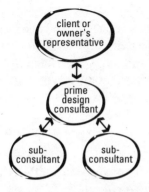

*Prime design firm-
coordinated project*

*Digital information site-
coordinated project*

protocol) sites for transferring files or project-based websites for collaboration online. It involves working out how to do this effectively using transfer speeds currently available, which at the time depended upon either modems limited to 33,000 or 56,000 baud, or paying for expensive permanent connections to the Internet. It also means thinking through coordination and control of project information.

With owner-coordinated projects, the owner(s) (or their representative) contract directly with design professionals. This can work if the project is professionally managed, but it also can result in poor communication between design professionals working on the project. This arrangement provides the owner with maximum control.

Another way to provide design services is for one firm to act as the prime consultant, hiring subconsultants to carry out specialized facets of the project. While this arrangement can improve coordination among design professionals, it sometimes results in weaker communication with the client, since all communication goes through the prime contractor. In this arrangement, the prime design consultant has the most control and responsibility.

A digital information site enables all parties—prime and subconsultants as well as the owners—to have direct access to the project information. Owners (or their project managers) could set up a project-based site if they wish to maintain primary control of a project. Or, prime design consultants could set up a project-based site carrying out the responsibility of coordinating and controlling a project. Service bureaus are emerging that provide project information sites for any party that wishes to set them up. For example, the Global Construction Network can provide a Project Specific Web Site™. Proprietary software we can lease, such as Blueline Online's ProjectNet™, does much the same thing. Organizational responsibilities could remain the same. What does change, however, is how information moves. Rather than being stuck in the owner's or prime design consultant's office, the project information could be accessible for all parties involved in the project. The project-based site enables anybody on the project team to access information online from anywhere, anytime, expediting transfer of information and smoothing out workflow.

Changes made to the source information are immediately available to all parties involved, avoiding the need to transfer multiple copies as well as avoiding the "snail mail lag" and the endless game of "phone tag" to let people know what is happening. Those participating can simply look in on the digital information site, see what is going on, and update what they are responsible for. The collaborative team needs to establish who has access to files and the authority to make

117

changes to the shared documentation. It is important to replace the "paper trail" (memos and transmittals that typically document the progress of the project) with a digital record. This can be done in "real time"—as transactions of information occur—saving each version of digital models—such as drawings, maps, images, even 3D models—as well as e-mail, and other correspondence that digitally document the progress of the project. We can search e-mail and other digital files by date, person, or topic quicker than we can search paper files if we needed to find out who did (or knew) what and when. Of course, these files need to be carefully backed up and protected as a record of the project. There is also the issue of signatures—or ways of authenticating key documents such as contracts—if we want to handle all the business and legal aspects online.

Standards

In the coordination meeting, Tony Palmisano addressed standards for drawings, information exchange, and documents. There also need to be some conventions for the way we use hardware and software. For example, he suggested using black screens and not white screens to select colors. When using AutoCAD, it is important to work with paper-space and model-space in a consistent way so others know how the drawing is put together and use "pack-and-go" to include all X-referenced files—as well as fonts and symbol files—when transmitting the drawing.

For drawing layer standards, Palmisano recommended using the AIA CAD Layer Guidelines, available from the AIA website. WLC also posted their office standards—following the (discipline, topic, subtopic) pattern of the AIA standards—on their website. It is important that all consultants work with layers consistently and apply only one color and one line type per layer to make it easier for others to turn information on, or off, as needed. Blocks should be drawn using layer guidelines and inserted on the 0 layer. Blocks should not be exploded. It also is important to use standard fonts so each party involved in the project has them, avoiding the need to send fonts with each drawing. WLC standardized their drawings using Arial—a TrueType Font. They also included their standard line and symbol sets at their website. Drawings should use origin points that relate to real-world reference points. This has advantages for site development and facilities management. The overall objective of the drawing standards is to ensure that all project team members can re-create and use the digital information shared online.

Information exchange standards involved agreeing upon service providers, magnetic media, and file compression. The service providers need to be MIME compliant with Internet access for e-mail and files as attachments, or using FTP, as well as provide access to the

World Wide Web. At the time, Palmisano recommended using Zip and Jazz disks or CDs for transferring large files and archiving, although he expressed interest in changing to DVD. PkZip and WinZip are standard compression software useful for reducing file size when transferring drawings online. Another strategy for reducing file size for transmissions is to send only the layers that have changed if the other parties already have the standard base drawing. Palmisano also strongly recommended that each team member maintain current virus protection software—such as McAfee or Norton—and immediately notify anyone who sends a virus that their system is infected so the team can protect the shared project information.

The documents standards Palmisano recommended reflected the software that the design teams typically used. For example, CADD documents used the AutoCAD.DWG format for the version the team was using or .DXF—the standard digital exchange format for drawings. For text documents he recommended using WordPerfect or MS Word. The RTF (Rich Text Format) was useful for moving between different versions of Microsoft Word as well as between Windows and Macintosh operating systems. For spreadsheets he recommended Excel or Lotus file formats; for graphics—JPEG, TIF, or EPS formats—compatible with Adobe PhotoShop and Illustrator; and for scheduling, MS Project. File-naming standards also are a key to sharing information online. Files must be carefully named in a consistent manner, preserving different versions of the file.

Project-Based Sites

Setting up a project-based FTP site involves establishing IP access to the information using passwords for security and file transfer protocols to move the information—preserving the format—so that others can continue to work on the file.

A project-based website uses a web page as a shell enabling others to use "plug-ins" such as DWF Whip. With these viewers, people can look at AutoCAD drawings online without using AutoCAD. The website can also create links between information both on the site and on the World Wide Web. Of course this makes it more difficult to upload files and requires having someone manage the site. It is important to use electronic transmittals when uploading files to either project-based FTP or websites to record who put what, where. Think routing and include transmittals stating changes. Manage changes to be cost effective and negotiate additional services for client-driven changes. Establish cooperation to maintain a high level of collaboration. Palmisano says, "Be nice, play nice. Consider the impact of changes on others."

Internet protocol (IP) Standards for transmitting e-mail and other files on the Internet.

plug-in An applet, such as Acrobat Viewer, that can work with internet browsers such as MS Internet Explorer or Netscape.

Examples

Here's a screenshot of WLC's current FTP site. Projects are arranged by job number, and all consultants have access to the projects they are working on, just as if they were project staff working in the office. From this FTP site project team members can download files they have authorization to access.

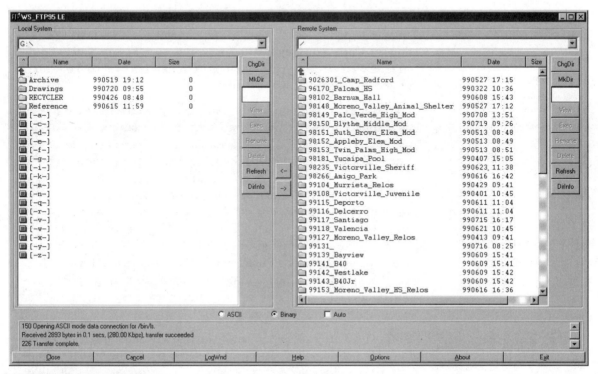

WLC FTP Site (Courtesy of Wolf Lang Christopher Architects)

Since running that coordination meeting and helping WLC set up their FTP site several years ago, Palmisano has moved on to another architectural firm, CHCG, in Pasadena, California. He wanted to have more opportunity to do architecture using digital tools and not just be a computer guru solving everyone's information technology problems—as he had become at WLC.

Digital tools are becoming available online to set up CADD sites for communication-aided design and data sharing. Java-based CAD viewer and a free service for posting and viewing CAD files are provided by Arnona Software (www.cadviewer). Document-centered markup tools are provided by C/Talk from ACS Software (www.acssoftware.com) and ReviewIT AEC from Cubus (www.cubus.net). Products for interacting on the web available from the ProjectCenter, (www.evolv.com), Instinctive's eRoom (www.instinctive.com) and Changepoint's Involv (www.involve.net) are just a sampling of what is emerging. Messaging combined with online document viewing and markup is available in NetMeeting that can be downloaded free from (www.microsoft.com) or included in the full installation of Microsoft Internet Explorer.

CADD and Workflow

Better access to information is changing workflow and how design professionals are able to carry out the design process. Palmisano observes, "With floppy disks, Zip drives and other forms of physical transport, we wind up with a faster version of the traditional design profession model of sending information back and forth. Interaction under this model tends to be limited to big meetings, and other discreet events." The current computer-aided design and drafting paradigm limits information transfer and communication. Meetings are expensive, time consuming, and often difficult to set up because of conflicting schedules and travel. Using a new paradigm for CADD—communication-aided design and data—we can shift to relatively inexpensive interactivity. Without high-speed connections, this is still limited to exchanging information, not sharing information.

Palmisano observes, "As the speed of exchange has increased, we have had glimpses of what this notion of sharing information will become for the design professions. One example is that it is now possible, with fast Internet connections, to keep all the information for a project in one location, central to all participants in the project. Instead of exchanging information about changing building or site conditions, changes are automatically included in all the participants' view of the project. It is as if all the team members—architects, landscape architects, engineers, owners, public agencies, etc.—are in the same room working together...in virtual workshop."

The development of communication-aided design and data is rapidly emerging, driven in part by the need to improve collaboration and the speed of disseminating information. It is becoming possible to readily share information stored anywhere on the World Wide Web as well as on a local area network. Significant advances in both software and hardware are addressing these needs. The recent AE Systems show in Los Angeles included many project-based information environment systems. Cable modems, digital service lines (DSL), and high-speed wireless transmissions will make it easier to use communication-aided design and data. The savings that can be derived from effectively using this information technology could quickly offset the associated costs. Eventually using digital information sites for CADD—communication-aided design and data—could become as pervasive as the present use of the phone and fax.

121

Summary
Guidelines for Setting Up Communication Aided Design and Data

1. **Build a team with the capabilities to share digital information.** This team should include people within your firm, or organization, as well as your consultants. Consider your clients to be part of the team. Also seek contractors who can implement projects by collaborating online.

2. **Establish standards for sharing digital information.** The standards should address drawings, information exchange, as well as documents standards and file naming conventions.

3. **Have the team begin developing these standards** by exchanging project files on disks, local area networks, and attached to the e-mail. Refine transfer strategies such as the use of compression software and upgrading communication software and hardware, if necessary, to handle the kind of information you wish to share online.

4. **Determine how to coordinate and control a shared digital information site for a project** you wish to carry out in this way. Resolve where the information will be stored and who will manage it.

5. **Establish a digital information site** for a project you select. Make sure the entire team commits to using it and set up a digital record with a person responsible for managing the site. Make sure the site is adequately protected and backed up so all parties have confidence in using it as the primary record.

6. **Manage the site to share digital information effectively** to get the maximum benefits from the investment. Make changes to improve the procedures so that the team's next project-based site is even better.

7. **Expand the use of the digital content by linking information to other web sites** where it may be used for other purposes such as fund raising or promotion.

Activities

Establish a Project-based Digital Information Site

There are tremendous possibilities for improving design collaboration, communication, and coordination by establishing communication-aided design and data sites. Develop your team's capabilities to do this using the guidelines listed here and select a project that would be a good candidate for establishing a project-based information site. Remember, it is essential to commit and use this site as the primary source and record for project information. Not committing will result in redundant efforts that can undermine the usefulness of computer-aided design and data sharing.

Manage the site and learn ways to improve it for your purposes. Cultivate your team's capabilities to use it effectively. Transfer what the team has learned to the next project, improving the return on this investment.

Look for ways to expand the use of the site. For example, is it useful to stakeholders? Is it useful during the review process? It may be necessary to transfer or link digital information from the project-base site to special websites set up for specific purposes—such as public review or fund raising. Tap in to the tremendous potential for using on-line information for design collaboration as well as for communicating with essential audiences and coordinating construction.

12

Applying Geographic Information Systems

Designers and planners, using spatial base information compiled digitally, can more readily analyze key considerations and make better decisions during planning, design, and management processes.

Organizations managing land and facilities need ways to inventory what they have and be able to optimize how they use it. Geographic information systems (GIS) provide tools for spatial analysis linking mapping with other forms of digital media and database management. In the past, setting up a GIS system required a considerable investment in hardware, software, and digital data. Unfortunately, it was too costly for most small communities, institutions, and businesses to set up and maintain. Today, that is changing.

Mark Sorensen set up Geographic Planning Collaborative, Inc. (GPC)— a spinoff from Environmental Systems Research Institute (ESRI), the geographic information software company that produces ARCH/INFO and ArcView. Sorensen applies geographic information systems and other digital tools to land planning and management dealing with urban and regional issues. Sorensen says. "We work to help our clients harness the power of these technologies, and to effectively integrate new capabilities into their projects and day-to-day operations." He is developing CityVision™ with the intent of helping communities use information more effectively. Sorensen envisions linking information generated by the assessors, engineers, facility managers, environmental health, and other services with more coordinated planning efforts to improve decisions made by city managers, councils and mayors who represent the interests of the people. He also is developing GIS QuickStart™ to help people get started working with spatial information in a way that is quicker, more cost effective, and technically more manageable. In these ways he can make the benefits of GIS available to local governmental jurisdictions, businesses, and consulting firms dealing with spatial information.

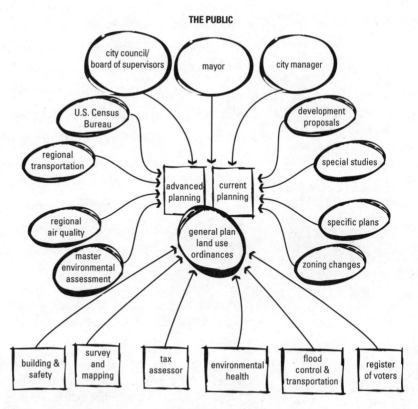

THE PUBLIC

Information food chain (adapted from Sorensen)

INFORMATION SOURCES

Sorensen's strategy is to help organizations get into GIS on an incremental basis by helping them identify long-range targets for what they would like to be able to do with GIS, and by providing a functioning system that immediately meets their most pressing needs. A QuickStart suite includes GIS software customized to meet the needs of the client, a bundle of appropriate data, and an implementation strategy to help organizations start up their initial GIS. This enables them to create visible and useful results right away while adopting a strategy for expanding and refining GIS over time. Sorensen provides continued assistance through onsite visits, or through online collaboration.

Some of the capabilities CityVision™ and GIS QuickStart for Engineers and Planners™ include are:

- Community demographic maps depicting present and future population distribution

- Detailed street maps, fully coded with addresses

- General plan and zoning information available in a digital form

- Property data linked to the thematic maps and reports

- Topographic information including slope, aspect, and elevation, as well as other land attributes

- 3D views and animated movement over the land form

The GIS Software Bundle with methods and tools that Sorensen packages includes:

- ArcView GIS software by ESRI. This desktop analysis and map presentation program enables creation of intelligent, dynamic maps using data from most sources across popular computing platforms. A user can work with maps, database tables, and business charts all in a single view.

- Village Atlas. A predefined ArcView project assists a community to access and navigate its information. The full functionality of ArcView is maintained so that more advanced users can also use the same software and information for custom queries, spatial analysis, and advanced map displays.

- Logical organization of thematic maps. This atlas-based approach organizes geographic data and in a logical and efficient manner. Groupings of thematic maps provide easy access as a starting point for organizing views. It is also completely open to customization and refinement.

- Standard symbols and display rules. GPS has developed a system of preset symbols to help users create readable and attractive maps. While these preset symbols provide a convenient starting point for most users, they are also open to customization.

Mark Sorensen compiles existing data and integrates it to help communities and organizations get started. This Data Bundle can include:

- Streets and street addresses. Up-to-date data available for the area, listing each street or highway segment, are ideal for geocoding purposes. This data can generate high-quality maps as well as sophisticated address-matching, routing, and allocation applications.

• Institutions and landmarks. This includes the location of government buildings, hospitals, churches, military installations, major retail
centers, universities, cemeteries, and natural features such as rivers
and lakes. Land use information provides a basis for spatial analysis.

• FEMA flood data. The Federal Emergency Management Agency
(FEMA) and the Flood Insurance Rate Maps (FIRMs) display zones of
potential flood risk. These maps help with flood plain management,
hazard analysis, and risk assessment.

• USGS basemaps. Raster or vector versions of the popular quadran-
gles, done by the United States Geological Survey (USGS), show land
features. Coordinating the 1:2400 topographic maps with a GIS system
provides a useful base for community planning.

• USGS Digital Elevation Models (DEM). Colorful maps showing
contours, slope, hill shade, elevation regimes, and aspect can all be
derived from the digital elevation models. This digital information is
useful when doing suitability or sensitivity models related to
landform. Coupled with ArcView Spatial Analyst, DEMs also provide
3D views and animated fly-throughs showing the terrain. This is very
useful for seeing what will be visible from different vantage points,
and for visualizing changes before they occur on the land.

• The census boundaries, demographic data, and projections. County,
tract, blocked group, and boundary information can be mapped at
high resolution. Mapping projections of key demographic variables
can incorporate current year's estimates, as well as five-year trends.

• Zip Code boundaries. The US Postal Service has established five-
digit Zip Codes. Mapping them can display thematic information,
especially related to e-commerce involving mail-order deliveries.

• Business locations and data. Locating businesses and places of
employment provide information for economic development. This can
identify traffic generators and other demands for infrastructure.

• Parcel boundaries, centroids, and tabular data. Maps of parcel
boundaries are often requested by property owners and developers.
Centriods, or points representing the approximate center of property
boundaries, can be tied to the Assessor Parcel Number (APN), address,
tract/block/Lot ID or other identifiers. Maps also can link to tabular
data on land parcels compiled by county assessor offices.

• General plan and zoning data. Local general plan and zoning maps
can be incorporated in a digital format with GIS. This makes it easier to
relate these plans to other information and update them, thus
responding to evolving needs.

127

Sorensen provides GIS Startup Plans to help organizations meet current needs while building an information infrastructure that will serve them in the future. For example, he helped the community of Cambria, California, compile information it needed to develop a water infrastructure ordinance while assembling information it could use for other purposes. Sorensen converted the existing PICK database to work with MS Access software on a PC. Updating existing parcel data with assessor parcel numbers (APN) and tract/block/lot (TBL) identifiers created a GIS database running on ArcView. Querying the GIS and Access databases generated information needed for the water ordinance. At the same time this was the beginning of a geographic information infrastructure that the city could use for mapping, facilities management, and planning purposes.

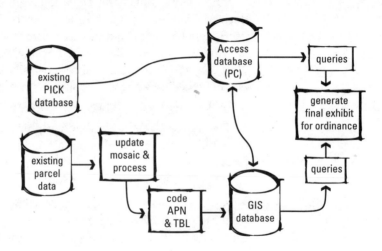

Applying QuickStart Methods to Cambria (Adapted from Sorensen)

Sorensen has been assisting planners, resource managers and others around the world in implementing GIS technology for over eighteen years. He recognizes that good planning requires good information and that information often exists in a variety of locations and formats. To help address this challenge, he has been assisting several countries in developing integration of spatial data across units of government. He also has been speaking at conferences around the world advocating building local, regional, and national geographic information infrastructure. He observes that GIS data could be the largest investment many organizations will make in developing their information system. He sees many overlapping interests and little coordination. There is tremendous redundancy and inefficiency in gathering data. Some data are incompatible. Sorensen is addressing the huge latent potential in sharing benefits and distributing costs to improve environmental planning and management by leveraging investments in GIS.

A key principal is recognizing that geographic information infrastructure (GII) is a fundamental component of sound environmental, economic and cultural development—much like highways, telecommunication networks, and educational facilities. Sorensen identifies the following key components needed to develop a global GII: institutional framework, spatial clearinghouse, metadata (the data about data), data standards, framework data, process standards, organizational standards, technology standards, communications, and partnerships.

metadata *Data about data. Used by clearing houses to help people find what information exists.*

Major initiatives are underway around the world. There is a European Spatial Metadata Infrastructure supported by the European Economic Union. There is a Permanent Committee on GIS Infrastructure for Asia and the Pacific, as well as the Australian Spatial Data Infrastructure. In the United States, a "National Spatial Data Initiative," called for in 1994, stated, "Geographic information is critical to promote economic development, improve our stewardship of natural resources, and protect the environment. Modern technology now permits improved acquisition, distribution, and utilization of the geographic (or geospatial) data and mapping. The National Performance Review recommended that the executive branch develop in cooperation with state, local, and tribal governments, and the private sector, a coordinated National Spatial Data Infrastructure (NSDI) to support public and private sector applications of geospatial data in such areas as transportation, community development, agriculture, emergency response, departmental management, and information technology." (United States Executive Order, 17671). The order further called for establishing a "National Geospatial Data Clearinghouse" for the development of a "distributed network of geospatial data producers, managers, and end users linked electronically." A Federal Geographic Data Committee (FGDC) was created to coordinate the federal government's development of the NSDI that federal agencies are required to use. This initiative is rapidly being adopted by state and local governments as well.

Private enterprise also is taking the initiative to address these needs. ESRI, the leader in GIS software, continues to advocate open standards accepting spatial data from a variety of sources. The ESRI website includes a "Data Hound" to help find geographic information online and is integrating an Internet map server with its GIS software. This will enable the user to access both local and Internet data using the same browser. ESRI also is working with Rand McNally and other partners to provide digital map information. Microsoft has established TerraServer, a website that has aerial photographs for many parts of the world (some of it using SPIN2 imagery purchased from Russia). Microsoft will mass-market MapPoint 2000, a GIS program that includes a detailed map of the United States along with census data. Autodesk now offers GIS software tapping into the considerable archives of AutoCAD drawings that now exist for engineering infrastructure, buildings, and site development.

There are five components to successfully implementing GIS. These include: hardware, software, data, procedures (for acquiring and relating geographic information to organizational processes), as well as people (the necessary staff training.) While many organizations recognize the need for a chief information officer (CIO), some are now beginning to recognize a need for a chief geographic information

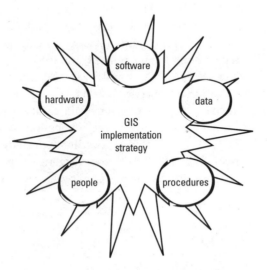

Five components of GIS
(adapted from ESRI)

officer (CGIO). Jack Dangermond, President of ESRI, asserts organizations could benefit from having CGIOs to develop their use of spatial information.

GIS programs are becoming more accessible. Wizards and focused applications can help design professionals access the power of GIS, while reducing the need for specialized training. Some applications of geographic information are embedded in special devices such as global positioning and navigation systems for airplanes, boats, trucks, and automobiles.

We are experiencing the rapid growth of digital spatial data. More information is becoming available and standard digital formats are making it more interchangeable between hardware platforms and software applications. For example, it is becoming relatively easy to move spatial information between CADD and GIS applications. The drawing and modeling capabilities of these applications are merging. There also is a movement to reference maps and air photos to global positions fitting them together seamlessly. This makes it easier for the user to find spatial information and gather only what is needed for one's particular area of interest. In addition, standards are emerging for resolution, accuracy, and data classification or coding schemes. Making this information available through the Internet is changing how design and planning professionals can find and acquire mapped information.

Who will become the "information architects" shaping cyberspace? Clearly we need people with an understanding of the processes that help people navigate and use information as well as an understanding of information technology itself. Universities need to educate professionals who can address all components of an information system.

Summary

Guidelines for Applying Geographic Information Systems

1. Determine the geographic information needs of entities you serve. Could they benefit from using a geographic information infrastructure (GII)? Should this infrastructure be set up internally, on a private basis, or linked to local, state, national, and even global GIIs?

2. Identify how to use geographic information systems (GIS) to address these needs. What software and hardware would be appropriate? Should it be set up as part of their operation, or out-sourced to consultants specialized in providing GIS services?

3. Select available software and hardware for these applications. Does the entity need a dedicated GIS with sufficient capacity to store and manipulate massive amounts of information? Can the entity work with desktop applications accessing and using information stored elsewhere on the World Wide Web? Does the entity simply need intelligent devices such as navigators in vehicles linked to spatial information?

4. Search for existing spatial data available online, on disk, or on tape. Coordinate with existing information generated within the organization or enterprise. Bundle information in ways that will serve their needs.

5. Identify additional sources of information that need to be compiled digitally to work with GIS. Develop strategies for acquiring this information. Can it be digitized directly, using remote sensing or other devices? Are there better ways to acquire it than scanning or hand-digitizing what may currently exist on paper?

6. Determine standards to integrate spatial information in ways that would be appropriate for the GII. Make sure that the information is correctly geo-referenced—tying it into global coordinates.

7. Identify potential sponsors for compiling this information as part of the GII sharing both costs and benefits.

8. Develop a concept paper addressing the five components of a GIS System and approach funding sources with a proposal to add this information to the GII.

9. Conduct a feasibility study to make sure that what you are proposing works. Then pursue implementation funding to acquire digital information that could be part of the GII serving the needs of individual entities as well as the needs of the larger community.

Activities

Use Geographic Information Systems for your Planning Project

Planners and designers need to compile base information digitally to facilitate more careful analysis of key considerations during planning, design, and management processes. The challenge is to select the appropriate information and compile it in ways that will be useful for community planning, and design management. GISs now link to global positioning, aerial photography and remote sensing, making it easier to acquire data. Standards are emerging for compiling this information and providing the metadata to make it more useful. This information can be compiled digitally as part of a GII and used in many ways with GIS programs. A GII could serve global, national, state, regional and community needs, as well as those of businesses and other organizations.

Design and planning professionals could be a catalyst for implementing the GII and compiling the necessary spatial information in appropriate digital formats. Working together we could help the GII grow globally, nationally, and locally, better serving the needs our clients and of society.

13

Involving Community Participation

Information technologies provide powerful tools for community participation. Using these digital tools effectively can help shape public opinion and assist public planning processes.

When I began this book, I had no intention of including a case study on the Cal Poly Pomona Campus Master Plan. The case study developed as I was working on this book—consuming considerable time and energy that I had hoped to devote to writing. Finally, I realized that what was happening on our campus provided wonderful examples of ways to use new media creatively as part of a community participation process.

The administration at Cal Poly Pomona launched an effort to develop a golf course on campus. This came about because of the need to raise funds using the considerable land resources that are part of the campus. The university was gaining more land from the closure of the Spadra Landfill and was in the process of updating the campus master plan. The golf course proposal began driving the master plan without careful consideration of the consequences.

The Golf Course Proposal

Most people in the university administration considered the golf course a "done deal" since they had spent several years and hundreds of thousands of dollars on the golf course proposal. Some faculty and students welcomed golf and the promises of generating badly needed funds. The Golf Course Steering Committee and Campus Planning Committee had rubber stamped this proposal. Without alternatives, they considered it inevitable.

As a professor in the College of Environmental Design, I could visualize better alternatives, and I had serious concerns about the golf course proposal. I began to find many colleagues who shared these concerns. In the spring of 1998, we began a year-long campaign, communicating through e-mail to turn this proposal around. Personally, I could not bear to see destruction of portions of the Walnut woodlands and understood the impacts of moving 1,500,000 cubic yards of soil from agricultural lands on campus to build golf greens on the landfill—an area of the campus we call LandLab. LandLab was originally intended as "A Laboratory for Education and Research in the Sustainable Use of Land Resources."

The golf course did not clearly support the mission of the university, the goals of the strategic plan, the original master plan for LandLab, and the approved EIR for the landfill. The golf course proposal also involved issuing millions of dollars of bonds, based on cost projections that did not include the total project costs. The market survey—used for projecting revenue—was seriously flawed. It failed to mention all the existing golf courses in the market area and did not take into account three new courses being proposed. Furthermore, not pursuing the golf course would avoid risks that could undermine educational programs and leave taxpayers with potential liabilities. Not developing the golf course also would avoid serious environmental problems that would be costly to mitigate or litigate and that would change the nature of the campus setting. The university is obligated to protect agricultural lands and the educational resources of natural lands that have been designated a "Significant Ecological Area" by Los Angeles County.

The university needed to focus on improving the campus master plan and to work together as a community to build upon shared visions and multidisciplinary collaboration. Not committing to the golf course would enable the university to pursue grant opportunities that could make it possible to use this land in other ways—addressing the larger issues of land reclamation, mitigation, preservation, sustainable agriculture, organic farming, regenerative design, renewable energy, and the ecology of commerce. The university could build upon its reputation for agriculture, regenerative design, and the business incubator being developed by our Extended University. A variety of possible uses for this 406-acre study area would offer more opportunities for "learning by doing" area than a golf course would provide. We needed to help our Campus Planning Office, consultants, and the Campus Planning Committee address the main issues related to building our campus community through an inclusive campus planning process involving the Academic Senate.

Leadership

As a senior tenured faculty member, I assumed a leadership role delineating faculty concerns in a memo to Bob Suzuki, the President of Cal Poly Pomona. It was important for the faculty to do this independently so that our Dean and department chairs in the College of Environmental Design could maintain normal relations with the administration. It seemed clear the administration did not want the golf course proposal challenged. I circulated a memo articulating concerns among the faculty. E-mail and word processing made this relatively easy to update by including feedback from faculty. A majority of the Environmental Design Faculty signed it. (Some faculty—up for promotion, or in lecture positions—shared our concerns, but felt too vulnerable to sign the memo.) After we submitted it to President Suzuki, he called a special meeting with the College of Environmental Design where he, and the administrative staff and their consultants, defended the golf course proposal. He did appoint a special committee of ENV faculty to advise on campus planning.

Advocacy

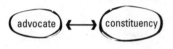

Advocacy

I was appointed to this advisory committee along with Rick Willson, Professor and Chair of Urban and Regional Planning, and Art Hacker, Professor of Architecture. The Academic Senate had become interested in this controversy and requested that we distribute the memo online. This opened up a campus-wide e-mail discussion of the campus master plan and golf course proposal.

Realizing that it would be necessary to help shape campuswide opinion, I began pursuing an advocacy role by writing weekly e-mail messages expressing concerns and distributing them to the campus community. Other faculty participated in sharing their concerns. Out of this came a growing awareness of land stewardship and sense of community. More people became aware of other opportunities for using this public land in ways that could serve the needs of both the campus and the community. The golf course was one of many issues that resulted in a faculty vote in of "no confidence" in President Suzuki. The widespread use of e-mail was instrumental in forming this widely held faculty opinion.

After the vote of no confidence, President Suzuki held a general meeting in the music auditorium with all faculty to discuss differences on issues—these ranged from faculty governance to the golf course. In the meeting, he conceded he was open to considering options to the golf course and invited us to submit proposals for further study.

The Campus Planning Advisory Committee met with the administration. Together, we determined that it would be helpful for the university to develop a more open process and continue to discuss campus planning issues further at the Fall Conference. Rick Willson was to prepare alternative visions and lead a discussion in the hope of helping the campus community arrive at a much-needed shared vision of the university. Art Hacker was to organize a workshop to help people explore forms based on different visions for the campus master plan and their implications. I was to explore alternatives for the 406-acre golf course study site. This site included the landfill as well as some of the campus agricultural and wild lands.

Consultancy

Consultancy

Meanwhile, I assembled a team of consultants and submitted a proposal for this study. The proposal was accepted by the administration and I continued as Coordinating Principal for the project. The team included: Jared Ikeda, a Lecturer in Landscape Architecture who served as Project Manager; Brooks Cavin, a Professor of Architecture; Curtis Clark, a Professor of Biology; Ted Humphrey, a Professor of English; and Mark Smith of Pario Research, an outside economic consultant. In addition, the team included staff from CEDG, Inc. (my professional firm that helped pull our study together). Because we were dispersing for the summer, it was important to have a team that could work online and prepare documents for submission at the Fall Conference. Together we produced a "Multi-Use Development Strategy" that Jared Ikeda presented at the Fall Conference while I was on sabbatical writing this book. We published the study on an intranet web site www.intranet.csupomona.edu/~muse/ so the campus community could access it and provide feedback. We also printed it in newspaper format for distribution on campus. Included with the publication was a questionnaire from which we hoped to gain further insight into which alternative uses were of interest to people on campus. We needed to find stakeholders for alternatives to succeed.

Political Activism

Political activism

After the fall conference, President Suzuki, in an effort to support faculty governance, determined the Academic Senate should decide on whether or not to pursue the golf course. The academic senate formed two committees: one committee in favor and one committee opposed. I became chairman of the opposition committee which also included Gwen Urey, Assistant Professor of Urban and Regional Planning; Sharyne Merritt, Professor of International Business and Marketing; Curtis Clark, Professor of Biology; and William Korthof, Student of Civil Engineering. Finding it difficult to arrange a common time and place to meet, we did most of our work collaboratively online. We had to prepare a five-page statement of arguments opposing the golf course and present it first to the steering committee and then to the entire senate for their vote. Since we worked online sharing working drafts, we also engaged other collaborators. These included members of the Multi-use Team, other faculty supporting the opposition, and students in the Black Walnut Alliance. It was helpful to have a varied group serve as a sounding board in developing arguments that might resonate with a broad cross-section of senators. The Academic Senate posted the final arguments at a website and sent out an e-mail message to inform the campus community where to find the site. They also sent the arguments to each faculty member as files attached to e-mail. The senators received printouts as part of their information package for the senate meeting. A lot of time and money had already been spent on the golf course proposal and it had gained considerable momentum. It was clear the opposition was coming from behind.

On February 17, 1999, both committees presented their arguments to the Academic Senate, in a session where there was standing room only. After a discussion involving the campus community, the senate voted to defeat the golf course—22 to 12. A vote "no" on the golf was a vote "yes" to:

- support established university policies

- avoid economic risk,

- preserve the environment,

- help heal our campus community,

- avoid public conflict, and

- provide more opportunities to meet the needs of the university and society.

Facilitation

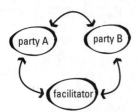

Facilitation

Stakeholders

Stakeholders

Information Technology

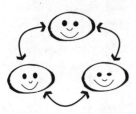

The loop

Hopefully, a shared vision will emerge for the campus master plan that the entire university can support. Possibly, the campus planning office or planning consultant can serve as a facilitator in helping the campus community sort out the best options. Or maybe the campus planning office will incorporate the faculty's recommendations. These need to reflect the values of the university community and be done in an inclusive way that involves potential stakeholders who can put plans into action.

From the Cal Poly Pomona experience, there are many lessons that we can apply to community participation in planning processes.

Each person who is part of a community is a stakeholder in that community. Communication technology provides new ways for stakeholders to interact and address issues that affect them. E-mail is a wonderful vehicle for community participation. Posting information on websites provides a quick way to publish it and get some immediate feedback. By developing models using digital tools (as described in this book), stakeholders can develop and explore shared visions. This can stimulate change, thus helping communities to work together to develop plans for improvement. Stakeholders breathe life into a plan.

One thing is clear—campus planning will be done differently in the future. Our sense of community is strengthened by information technology. Now, it is possible to not only disseminate information online, but to make it part of a more democratic decision making process.

A GIS Center at Cal Poly Pomona is now compiling a geographic information infrastructure for the campus. It is hoped that this information and information the Physical Plant Office has already been compiling can be combined and will be used to inform the campus planning process.

The campus master plan can become a vehicle for teaching and learning, as each discipline uses the realities of the plan as a supremely relevant case study for "learning by doing." Sharing information online can enriched these learning experiences. Sharing the results of the case studies can inform the decision-making process.The master plan needs to consider carefully earthquake faults, groundwater resources, and agricultural soil.

We found our e-mail campaign most effective when it focused on the issues with honest and credible statements. Insofar as we avoided personal attacks, we were able to address problems with minimal polarization. A tricky part of e-mail was the handling of distribution lists. By effectively managing distribution lists and carefully crafting our messages, we were able to reach the intended audience in a positive way.

Spamming (sending messages indiscriminately) only hurts a cause, aside from being a nuisance that undermines the usefulness of e-mail. At Cal Poly Pomona, most faculty seemed to appreciate the messages we sent regarding the campus master plan; there were a few who wanted to lighten the load in their e-mail box, particularly over the summer. It was even more important, however, to include all those who needed the information—so they didn't feel left out of the loop. The most serious misunderstanding I encountered was when a person was inadvertently not included in some of our distributions. It took me awhile to realize whom the Academic Senate distribution lists I was using did not reach. While online collaboration through e-mail became a very effective vehicle for gathering support and moving the cause forward, it did not supercede personal meetings or casual conversations on campus. There is no virtual substitute for a real community, but online interaction can help communication and enhance a sense of community.

Websites provided the opportunity to share more extensive information including maps and graphic images. Ideally these sites became a vehicle for two-way communication—sharing information and getting information back. Desktop publishing is less interactive, but may be a more effective way of getting more information out to an audience. We did include a questionnaire at both the website and in the publication we distributed about the multiuse development strategy. The real challenge is to find stakeholders who would commit to the endeavor and create an environment where they could succeed. Sharing information online provided ways to help nurture these efforts.

Organizational Structure

There are different types of organizations and it is important to develop a sense of how they work. Organizations—like a university—may operate in several different ways.

"Top down" organizations require administrative control. They are hierarchical and have a "chain of command." Orders emanate from the top to be carried out by underlings. Funding from the top needs to be carefully accounted for. We see this autocratic structure in military organizations and in some corporations. It is the thrust of the cartoon character Dilbert. These organizations may attempt to use electronic media for control, monitoring activities and looking in to correspondence, but incriminating e-mail can also be used against organizations, as has happened in court.

head

Top down

140

people / markets

Networked

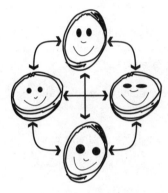

Collaborative

"The more I think about the Internet the more I realize that it is much bigger than the wires and computers that seemingly house it. It is in some ways like millions of doors that millions of imaginations flow through. Dreams, desires, ideas, hopes, emotions, language and expression spark and ignite across its vast and expanding surface."

— Gerry McGovern
gerry@nua.ie

"Networked" organizations serve people or markets. They involve democratic governance, carrying out initiatives, or responding to market demands. E-mail can be used very effectively in these organizations to provide necessary feedback, help shape opinion, and rally support.

"Collaborative" organizations are mission - or project-based. They involve stakeholders. This organization becomes a forum in which to operate; yet, there is autonomy within that framework. Information technology provides a vehicle for collaboration. It stimulates ways for groups to interact—sharing minds, responsibilities, benefits, and information. Digital tools provide opportunities to quickly establish a virtual collaborative organization for a project or mission.

It is likely that information technology will provide a vehicle for shifts in organizational structure. Providing they have communication skills, digital tools can help more people "take part" in decision-making processes.

Gerry McGovern who writes "New Thinking"—a weekly e-mail message that I subscribe to—has written a book called *The Caring Economy—Digital Age Business Principles* in which he develops a new philosophy for human interaction online. "It is a book about how we all need new attitudes, new rules and new business principles for success in a digital age economy and society."

Summary

Guidelines for Stimulating Community Participation

If used effectively, communication technology can help nurture community participation. To do this we need to do the following:

1. **Determine what audience to reach.** There are many parties to any problem. Each needs to be reached in different ways. For example—within a campus community—there are administrators, consultants, faculty and staff, students, and visitors. There is also the neighboring community and, in the case of a public institution, the taxpayers who own the public land.

2. **Determine what role(s) to fulfill.**

- **leader**
Although it is often tempting to abdicate leadership given the pressure of organizational life, social responsibility may demand assuming it some situations.

- **advocate**
Advocates need a constituency, but information technology can help build that.

- **consultant**
Consultants need a client; information technology helps consultants work more collaboratively.

- **political activist**
Political activists need to influence the decision-making process, shaping opinion and getting out the vote. Again information technology can help.

- **facilitator**
Facilitators need to be neutral, seeking creative solutions that may result in win-win situations or alternatives that would involve acceptable compromises.

3. **Understand the decision-making process.** Each organizational structure has a different culture and its own set of rules. Only by understanding those rules can you effect change if necessary.

4. **Clearly package information in ways that have bearing on the decisions.** Some people base decisions on moral or ethical values. Others need logical arguments. Still others respond most favorably to emotional appeals. Realizing that people make decisions in many different ways, we can use new media to communicate more effectively.

5. **Work toward a shared vision engaging stakeholders.** Look for win-win situations or at least for compromises that are mutually acceptable. Nurture relationships with stakeholders who can breathe life into a project, or plan.

142

Activities

Engage Community Participation

Consider the organizational structures you are working with. Are there more effective ways to engage community participation utilizing the information technology that is available?

Select a situation that needs public involvement. Can you provide leadership to help shape public opinion? Can you visualize change? Is there a constituency you can serve as an advocate, drawing upon your capabilities as a design professional. As a consultant, how can you engage community participation in a public project, help educate the public, and nurture stakeholders? How can prospective users evaluate designs and help refine them so they have a better chance of approval or success in the marketplace? Participate in the political decision-making processes focusing more thought on design considerations that can regenerate and sustain the environment. In some situations, design professionals may be able to serve as facilitators finding win-win solutions where there are controversies.

Consider what information is needed to make good decisions. Develop models that represent key realities. Visualize change by illustrating the alternatives using a variety of models. Share this information with other stakeholders and encourage people to work collaboratively.

Use the guidelines provided here to shape your design practice and work with information technology to serve the public good

Strategies

"Realizing the Potential"

PART I   Methods

Design teams—including managers, design professionals, and staff—
need ways of working effectively. This part of the book offers methods
to use new media and digital tools creatively and productively for
design communication in professional practice. It can be a helpful text
in courses on design methods and new media for design communication, as well as computer courses that go beyond application training.

PART II   Case Studies

Good examples demonstrate how progressive design professionals are
using new media and applying digital tools in their practices. The case
studies examine a range of design practices—writing and graphic
design, creating multimedia and special effects, landscape architecture,
architecture, and planning. The case studies explore what works in
office environments as well as in virtual offices and electronic studios
using online collaboration.

PART III Strategies

Once we invest and commit to using new media and digital tools, it is
important to find ways to get the most out of them. This part of the
book is devoted to realizing the potential inherent in information technology. It will provide valuable help as your design team addresses
change.

APPENDICES

14

Regenerate Body, Mind, and Spirit

Design professionals can find ways to regenerate body, mind, and spirit while working creatively with new media. We can nurture our physical, mental, and even spiritual capacities to help us meet the challenges we are facing. Rapid changes affect the way we practice. Environmental degradation presents moral dilemmas that require soul searching. As design professionals, we have the capacity to improve the environment in many ways.

Digital tools offer freedom to work anywhere, anytime, which tempts us to try working everywhere all the time. We need to carefully assess how we use these tools and learn to regenerate—both ourselves and the environment. The repetitive motion of clicking a mouse and keyboarding can cause the triggering muscles to tighten. Working without sufficient rest and under the pressure of deadlines may result in muscle knots that can pinch nerves causing fingers to become numb. Unfortunately, I experienced this while writing this book, which is rather ironic because I advocate using digital tools. How do we address this?

Balance

human regeneration

rhythm

computer image regeneration

Interactive rhythm

The key is balance. We need to balance work with rest even though our tools are inviting and accessible. A balanced nutritious diet is important. Plenty of water helps lubricate muscles. Certain foods, such as pineapple and bananas, contain nutrients that are good for muscle, tendon, and nerve tissues. Supplements, such as the vitamin B complex and others, also stimulate healing. Promoting circulation through exercise and heat helps nourish muscle tissues, although, initially, cold can help reduce swelling. But once muscles have contracted in knots, it may take bodywork, such as massage or acupuncture, to release the tension. I have found yoga to be extremely helpful for healing and maintaining a holistic balance. Yoga enables us to focus and release muscles by directing the energy of our breathing and blood circulation. Through a large repertoire of asanas (postures) we can stretch in many beneficial and pleasurable ways. Yoga also helps us maintain more balanced posture.

We also can focus energy to regenerate wholeness within individuals as well as well as among groups and large organizations. Chi Gung is an ancient Eastern practice that can help us access our internal energy. More recently, Stèphano Sabetti has developed a Life Energy Process®

regenerate *Natural process to form again or renew. To be spiritually reborn. For example, while computers regenerate images on a screen, human minds recenter and refocus — regenerating mental images in rhythm with life's energy.*

that can help us access and sustain the flow of energy of our centered being. His book, *The Wholeness Principle*, describes this approach that is now being taught through the Institute for Life Energy.

Simply doing some exercises while working and taking frequent breaks can prevent injury. It can also help regenerate life's energy. If we look at the rhythm of our own thought processes, we realize there is also a need for human *regeneration*. Physically, mentally, and even spiritually we cannot be constantly on "go." We need time to contemplate—time for "so." Ideally, we can relate our own rhythms to the rhythms of the computer. We can learn how to pace ourselves, relating our regeneration time to the computer's regeneration time. In other words, while the computer is regenerating the screen or doing other tasks, like saving, sending, or printing files, use the time to regenerate your own body, mind, and spirit.

Relaxing

Obviously, we should limit work sessions on computers to what we can do comfortably. However, there are ways we can relax our bodies that enable us to maintain peak performance for a longer time. Deep breathing is a good way to begin.

1. Breathing Deeply

Breathe out

People usually breathe very shallowly when using a computer. We may compound this by slouching in a poorly designed chair, or possibly hunching over to look at a monitor positioned too low. Poor posture collapses the chest cavity, reducing the volume of the lungs. The act of breathing provides oxygen to burn the energy we consume. Therefore, if we breathe shallowly for a long time, we begin to feel logy. Our minds become dim, and our spirits may be low. We will probably find ourselves involuntarily stretching and yawning when our body senses oxygen depletion. We can consciously learn breathing exercises that will help keep us alert. Try doing these exercises during your regeneration time.

Breathe in

A simple breathing exercise begins with exhaling to a count of 5 pressing your navel toward your spine while rolling your shoulders forward to collapse your chest. Inhale to another count of 5, filling your whole abdominal and chest cavity with air by stretching your arms back to expand your chest. Even yawn if you need to. Pause briefly, and then repeat—slowly exhaling and inhaling—until you feel you are completely alert.

Notice as you do this breathing exercise that your posture will naturally improve. Maintain good posture to sustain good breathing.

2. Rolling the Spine and Balancing the Body

Good posture is a key to staying alert. Improve your posture by rolling your spine into straighter alignment, starting at the base of your back, then slowly working up to the area between your shoulders and finally stretching your neck by gently pointing your chin to the ceiling. Think how gracefully a cat stretches its body. Enjoy the sensation of your muscles relaxing, causing your blood to circulate freely and energize your body. Remember the feeling of having your body balancing easily on the base of your spine and retain this posture. Properly position your chair, keyboard, and monitor so you can easily maintain this feeling. Also make sure your hands have a place to rest so your neck and shoulders can remain relaxed. It is especially important that your wrists are straight when using a keyboard. Repetitive finger strokes through bent wrists may damage the carpal tunnels through which your tendons move.

Roll spine

3. Looking Beyond the Monitor

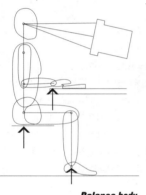

We use our eyes intensely when working with computers. Poor contrast, poor resolution, flicker, and glare from a monitor compound eyestrain. Therefore, it is not surprising after working with a computer for awhile, that our eyes can become tired. We may begin to lose focus or feel spasms.

You can resolve some of this by carefully adjusting your monitor. Also control external light sources to reduce glare. Appropriate eyeglasses can correct focusing problems. Beyond that, you should limit your work sessions to what you can do without straining your eyes. Some simple eye relaxation exercises can help rest your eyes as you work. Again, do these exercises during regeneration time.

Balance body

Each time your computer is regenerating or doing some other task, get into the habit of fixing your eyes on something in the distance, causing them to refocus. When working with a very fast computer that takes little time to regenerate, look away from the monitor periodically. This is easiest to do if you are in a room that is spacious, offering longer views. You can also do this by looking out a window. However, make sure the light level of your monitor and that of the outside view are similar, so you don't suffer glare when changing your view. Even without a long view available, you can change your focus by imagining that you are looking at something far beyond the monitor. You will notice the monitor go out of focus when you do this. Using your peripheral vision, you can perceive when the regeneration or other task has stopped. Then, if your eyes feel ready, you can refocus on the monitor and proceed with your work.

Look beyond monitor

4. Relaxing Eyes and Ears by Palming

Palm eyes

Palming is another excellent way to relax your eyes. Bring your head to your knees and cover your eyes with the palms of your hands. Relax the muscles in your neck and back. Keep your eyes open so you are sure not to press your eyeballs. Peering into darkness, move your eyes slowly and rhythmically from side to side and then in circles. Doing this, you will notice that the afterglow of light and color disappears. Without the external stimulus, your eyes soon relax. (Eventually, light and even colors may reappear once you are able to focus deeply within yourself, drawing upon your inner energy.) The warmth of your hands and the movement of your eyes stimulate blood circulation in this crucial area. Make sure you have a proper diet, since eyes are high consumers of vitamins. Because palming takes a little longer than refocusing, you might get in the habit of doing this while you are saving large documents or sending or printing files.

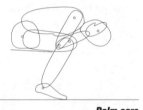

Palm ears

Try palming or cupping your ears. Cover your ears with the palms of your hands. Do it firmly, but be careful not to cause pressure or suction on your eardrums. You will find that external sounds subside. Suddenly, you notice the sounds inside your head. Lower your head to your knees and close your eyes. Rest there, relaxing the muscles in your neck and back, listening to the sounds in your head. This silence gives your ears a chance to rest. Also the warmth of your hands and the lowering of your head cause blood to flow to your ears, bringing renewed energy. This can also release tension in your neck and back. Do this while the computer is printing. It is a good way to avoid noise. Notice you can focus on breathing when palming your ears.

By now you realize these exercises may be most suited for the privacy of your study. Some of the exercises such as breathing and changing the focus of your eyes are subtle enough to do among strangers. On the other hand, some of the exercises, if done among strangers, can make you the center of unwanted attention. If possible, create your own support group where the people you work with understand the usefulness of these exercises.

Certain muscle groups can become tense because of the repetitive actions typical of using keyboards and pointing devices. These usually are in the areas of the neck, back, shoulders, arms, hands, and fingers. We can relax and stretch these areas at frequent intervals while seated at our computer. Here are some exercises you can do to regenerate while waiting for the computer to regenerate and carry out its tasks.

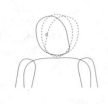

Roll neck

5. Rolling Your Neck

Relaxing

Rub your neck. Make it warm.

Stretching

Roll your head slowly from side to side. Relax your neck, letting the weight of your head stretch the muscles in your neck and back. (You may notice a crunching sound. This is normal.)

6. Rolling Your Shoulders

Relaxing

Rub your shoulders down to your elbows. Make your shoulders and upper arms warm.

Stretching

Slowly roll your shoulders forward and back. Push your chest out as you roll your shoulders back. Then reach your arms toward the ceiling and alternately stretch each side of your torso by extending your hands as if you were picking fruit from a high branch.

Roll shoulders

7. Rotating Your Arms, and Shaking Your Wrists

Arms and Wrists

Relaxing

Rub your forearms. Massage your wrists by gently twisting and kneading them with you hands.

Stretching

Move your hands in a circular motion, gently stretching your wrists. Then hang your arms limply at your sides and shake out your wrists by rotating your arms.

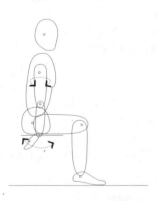

**Rotate arms
Shake Wrists**

8. Opening Your Hands, Stretching Your Fingers

Hands

Relaxing

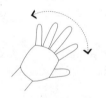

Open hand

Rub and knead each hand with your opposite thumb, fingers, and palm. Work your thumb around your palm. Then run your thumb up the valleys on the back of your hand between the bones connected to your fingers. Rub in the direction of your heart.

Stretching

Open your hands, spreading your fingers as wide as you can. Repeat this slowly several times.

Fingers

Relaxing

Stretch fingers

Use the thumb from your opposite hand to gently rub your fingers toward your heart. Make sure you rub each finger and thumb.

Stretching

Clasp your fingers together and stretch by gently turning your hands inside out as you sometimes see a pianist do before a concert.

When doing these exercises, do not hold your breath or force any of the stretches. Stop if you feel any pain. The stretching sensation is actually quite pleasant. Take time to enjoy it. Think of yourself as a concert pianist warming up at the keyboard as you gracefully move from one exercise to another. Let your "music" brgin by relasing your internal energy through whatever application you are working with.

Concert pianist

Releasing

These exercises can bring a feeling vitality into your body and mind. Because electronic media can become so absorbing, make sure you occasionally turn attention to yourself and determine how you are doing. If you can't regenerate a feeling of vitality using these exercises, it is time to take a break.

Consider fitness part of your daily routine. Relaxation and fitness are like sleep and good nutrition. We don't function very well without them.

Relaxation techniques found in yoga and similar practices help us learn to relax and release our minds.

Letting Go of Our Bodies

Here is a mental relaxation exercise that involves focusing on breathing and letting go of our bodies. Begin by taking two or three deep breaths, exhaling to the bottom of your lungs each time. While sitting, you can bring about controlled relaxation by sequentially letting go of each part your body, feeling the forces of gravity, as opposed to letting your whole body go limp like a rag. For example, start by releasing your legs, allowing them to rest motionlessly on the floor. Next release your lower back, putting all the muscles in that area of your body at ease. After that, release your torso by carefully balancing on your spine. Release your neck by carefully balancing your head on your body. Release your arms, allowing them to hang at your sides. Then relax your jaw—even letting your mouth hang slightly open. Next turn your attention to your eyes—which you can close gently. Then focus on your forehead, which you should put at ease without furrows. Similarly releasing your scalp and ears may cause them to tingle because of the added blood flow. Your body will provide clues to indicate you are relaxing. Your limbs will feel heavy, and you will experience a sinking sensation as your body responds to gravity. You will naturally begin to balance yourself with good posture if you are sitting or standing. You will feel a warm sensation as your blood circulation improves. Your breathing will become deeper and more rhythmic—as it does when you are beginning to fall asleep.

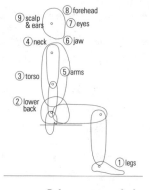

Release progressively

Releasing Our Minds

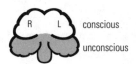

conscious

unconscious

Inner realm

Mastering: combine inner space and cyberspace

Releasing our minds is important for relaxing and being able to work at peak performance for longer periods. This can also help us explore our inner spaces so we can learn to relate our creative drive to what we do in cyberspace.

Most cultures have traditions related to meditation. Certainly major religions—Buddhism, Christianity, Hinduism, Islam, Judaism, and Taoism—involve this type of experience. Religions have developed rituals for meditation, be they simple prayers or mystical experiences. Aside from the religious implications (which vary), the common theme is learning to access deeper levels of consciousness and becoming a wayfarer in our inner spaces.

Unconscious levels of our minds are the foundation of our values and the source of our motivation. Mental energy derived from these levels is what fuels our creative drive. We can explore our unconscious; there are many ways of doing this. Creative thinking skills relate to unconscious as well as conscious mental energy. We can learn approaches to exploring our own inner space so we can transfer energy from our unconscious to images we can work with consciously. This enables us to express our insights so we can work with them interactively, using digital tools to share them with others through cyberspace.

Interactive multimedia enables us to participate consciously and collaboratively in imaginative experiences. We can use digital tools to actively explore virtual reality, engaging our imaginations. Drawing upon our feelings, thoughts, and emotions, we can give expression to designs in virtual reality that can be implemented in the real world. Information technology provides a vehicle for getting back and forth between our imaginations (and what we can express in virtual reality) and the situations we deal with in the real world.

Today's information technology may help transfer the energy of involution—which descends from cosmic realms—and the energy of evolution—the progressive opening to wholeness—in ways that promote healing within individuals as well as cultures.

Imagination

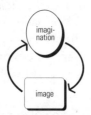

Active imagination

We can use our imaginations very simply and quickly, or much more elaborately over a longer period of time. As you probably realize, however, it is not always easy to use digital tools to develop quick and simple representations of what you experience, drawing from unconscious thinking. Sometimes it is easier to start with simple mental notes, verbal statements, or hand-drawn sketches. Eventually you can do this electronically, once you are comfortable using digital tools. The advantage of using computers is that you can work more interactively with increasingly elaborate models of what you imagine. Evaluate how you are recording your insights. Make sure you find approaches that are both comfortable and interactive. Develop strategies for communicating your insights and transferring ideas to objects you create.

Robert Johnson, in *Inner Work*, provides a four-step procedure for what he calls "active imagination." Active imagination involves an interaction or dialogue with your unconscious levels that results in some action.

Here is how to apply Johnson's four-step approach for active imagination using new media:

1. **Invite the unconscious.** Do this by contemplating or reflecting upon some stimulating feelings, ideas, or concepts.

2. **Dialogue and experience.** Use digital tools to work interactively with electronic media to develop a digital journal. For example, write, draw, or build digital models representing what you are imagining. Representing this emerging reality that you experience interactively can further stimulate your unconscious mind. Become conscious of your unconscious by recording your insights. Record thoughts in ways you, and others, can interact with.

3. **Add the ethical element of values.** What insights have value relative to the problems at hand? Are they practical? Are they feasible? And, ultimately, are they ethical?

4. **Make insights concrete with physical ritual.** Create an image that will help you visualize your insight more clearly. For example, incorporate this vision into what you are talking about, writing, drawing, or modeling. Refine your insights and incorporate them into a virtual reality of what you are creating so you and others can experience it more thoroughly. Enjoy the ritual of sharing what you are imagining with others.

Dreams

We access our unconscious through dreams. This is something that we all experience regularly. Working with our dreams can be very helpful to our mental health and personal growth. Since dreaming is an aspect of creativity, it can also be useful for developing our creative capacity.

You might find special places where you can do this, such as a couch in your study, a secluded cabin, or even a comfortable place outside—perhaps under the proverbial apple tree. Immerse yourself in a project before you go to sleep. Contemplate it as you are dozing off. You might keep a sketchpad, recorder, or notebook computer nearby so you can take notes that will help you remember insights you derive from your contemplation. After you have slept, take the time to reflect on what you were dreaming about. Again use a sketchpad, recorder, or computer to take notes that will help you remember your dreams. Do this even if the dream doesn't seem related to the project you were contemplating when you went to sleep.

Use Johnson's four-step approach for interpreting your dreams:

1. Make associations. How might the dream relate to what you are contemplating (or to anything else in your life that matters)?

2. Connect dream images to inner dynamics. Does the dream reflect insights that concern you?

3. Interpret. What conscious feelings or ideas can you derive from the dream?

4. Carry out rituals to make the dream concrete. Can you create an image of your dream that will help you visualize it more clearly? Again, can you incorporate it into what you are writing, drawing, saying, or modeling? Can you use it in models you are developing in virtual reality? Can you act upon your dream to experience it in reality?

You may find that you already do some of this naturally. Probably the most difficult thing is taking the time, finding the right place, and relaxing—getting in the right mood. Notice the times, places and thought patterns that help you do dream work. Get in the habit of incorporating these patterns into your daily life. Also notice that stimulants like coffee, alcohol, tobacco, or drugs are counterproductive. They may overstimulate your mind or thrust you into a stupor, thus making it much more difficult to derive insights from your dreams.

We can explore our inner spaces and transfer some of what we discover to cyberspace. By doing this we connect with our imaginations, or dreams, and can work with insights more consciously by interpreting, modeling, and making our insights real, using a variety of digital tools. Shaping virtual realities by using digital tools can help provide ways to create designs that resonate with meaning. We can draw upon the creative energy in the unconscious levels of our minds by learning to use mental imagery. This "art spirit" can regenerate creative energy in ourselves and others through creative collaboration.

Summary
Strategies for Regenerating Body, Mind and Spirit

1. **Balance your activities.** Even though you may have access to digital tools anywhere, anytime, try not to use them everywhere all the time. Make sure you get rest, balanced nutrition, and exercise.

2. **Relax your body.** While you are working at a computer, take a few moments to occasionally do the following exercises: breathe deeply, roll your spine and balance your body, look beyond the monitor, relax your eyes and ears by palming, roll your neck, roll your shoulders, roll your arms, and shake out your wrists, open your hands, and stretch your fingers.

3. **Let your body go** by feeling the forces of gravity pulling you progressively from your feet up through your lower back, balancing your spine, neck and head, releasing your jaw, eyes, forehead, scalp, and ears. Let your breathing become deep and rhythmic.

4. **Release your mind** by turning inward and meditating, accessing deeper levels of consciousness and becoming a wayfarer in your inner space.

5. **Use your imagination actively to explore insights** by inviting the unconscious. Converse with and experience your insights by interactively using digital tools to develop a digital journal. Add the ethical element of values to evaluate your insights. Make the insights concrete through a physical ritual of incorporating this vision into what you are talking about, writing about, drawing, or modeling—as well as refining in virtual reality—and enjoy sharing what you imagine with others.

6. **Draw upon your dreams** by making associations, connecting dream images to your inner dynamics by reflecting on them, interpreting your dreams to derive feelings and ideas, and carrying out rituals to make a dream concrete—incorporating it into virtual reality or acting upon it in reality.

7. **Get in the habit** of incorporating these patterns into your daily life. Connect your inner space with cyberspace and cultivating an "art spirit" to regenerate creative energy.

Activities

Engage Your Imagination and Draw Upon Deeper Levels of Consciousness

Design professionals can work creatively using digital tools to develop our insights by drawing upon deeper levels of consciousness. Doing this helps us regenerate body, mind, and spirit and maintain human balance that avoids burnout.

Previous activities have involved digital design journals used mostly to record sensory perception and gather information that we can acquire online. Now use this journal to draw upon and nurture your inner strength. Access and explore your inner self by using the approaches and strategies identified here to relax and let your body go, releasing your mind, and using your imagination actively to explore insights and dreams

Find a time of day and place where you enjoy meditating. Get in the habit of meditating early in the morning when you wake up, or some special time during the day, or in the evening before you go to bed. You may do this in a favorite chair, or in a study, or while taking a walk, or actually anywhere—even on a plane or train while you are traveling.

Meditate where you have a sketchpad you can use to record key words and diagrams you can later develop using digital tools. Or better yet, use digital tools to directly record and interact with your insights. These digital tools may simply be a recorder you can speak into, or a personal digital assistant with a stylus for diagramming and sketching, or a notebook computer with a word processor and graphic programs. Reflect upon design issues at hand—projects you are involved in. Use your imagination actively and also draw upon your dreams.

Leonardo da Vinci, experimented with sleeping for short periods so he had more time to pursue his creative endeavors. Also, by sleeping and waking more frequently, his mind went between dream and conscious states more often. The inventions he dreamed about were hundreds of years ahead of their time.

Interact with your insights and enjoy developing them. When you are comfortable with the ideas you have generated, share them with others.

Avoid impulsively imposing them on others, but look for the right time when people are receptive to share them. Nurture creative collaboration with others who share your concerns. Connect online as well as in person. Digital tools provide us with new ways to explore and share our insights.

15

Interact and Collaborate Online

Design teams that interact effectively online can learn to correspond and collaborate more quickly and easily. As design professionals, we need to learn to interface with digital tools that can help us form teams to implement projects in a more timely, energy efficient and professionally empowering manner. Digital offices evolve as we make better use of digital tools. For any project or endeavor we can even establish a virtual office, or team, that is location-independent, but linked through the Internet. There are basic patterns for doing this regardless of what enterprises we are involved in—community planning and resource management, communication design and entertainment, architecture, engineering and construction or product design. These patterns are directly related to the manner with which our bodies, minds, and spirits interact with digital tools, emerging information environments, and our fellow collaborators. The Internet and World Wide Web are, in many ways, well suited to the dispersed and multi-faceted nature of these enterprises and offer a growing assortment of digital tools for design professionals to use. The key to interacting online, beyond acquiring access, is to become comfortable with using digital tools. To collaborate on-line, we need to use shared information environments and develop digital models that represent the realities we are working with.

digital office *An organization that makes extensive use of digital tools.*

virtual office *A location-independent organization that links team members through the Internet, Intranets, and Extranets.*

A Web Workstyle and Lifestyle

Design professionals can adopt a Web workstyle by getting in the habit of accessing on-line information, to quickly gain input. By communicating online we can expand our capability to collaborate. Design professionals can develop a presence on the Web providing visibility for our projects and talents and linking to those who have needs for our services. We can cultivate professional relationships and build teams for online collaboration. A Web workstyle can enable design professionals to work where we can relax, think, and avoid rush hour commuting by transmitting information and ideas electronically from our homes and studios or many other places we care to be.

Aside from a Web workstyle, adopting a Web lifestyle connects us to family and friends, as well as to news that interests us, and a growing electronic market place where we can quickly research, select, and order what we want with a few clicks of the mouse. Better communication technology enables us to spend more time in our homes and communities, enjoying local pleasures like walking and bicycling, while being able to stay in touch with people and events all over the world.

When we are traveling to enjoy other places, communication technology enables us to stay in touch with family, friends, and colleagues at home. I have been fortunate to travel for extended periods of time while being able to correspond with friends as well as work collaboratively with colleagues, using a notebook computer with a modem. Through the Internet, my wife and I stay in contact with our daughter and son-in-law who live in Italy. We even enjoy sending pictures back and forth. As the bandwidth of this connection becomes broader, I look forward to exchanging video clips to share events in our lives, even though we are thousands of miles apart.

The Interface

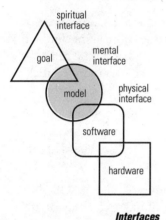

Interfaces

It has been observed that while laborers may work with their hands, knowledge workers work with their heads. Crafts people may work with their hands and heads, while artists (as well as design professionals) work with their hands, heads, and hearts.

Similarly interacting with a computer interface involves our body, mind, and spirit, and, as with art, there are different levels of involvement. To work with digital tools creatively, we need to find ways to bond with them. At one level there is the physical interface—which relates to the way we work with the hardware devices. Understanding physical human factors and ergonomics can help prevent repetitive motion problems. At another level there is the mental interface, which involves software procedures and thinking skills. At still another level is the spiritual interface, having to do with a sense of attachment, empowerment, and meaning. The nature of this level is harder to articulate, although this is an important part of our being. Each level influences how we interact with digital tools and new media.

By learning to work with interfaces physically, mentally, and spiritually, we can apply familiar user interfaces in new ways and also be open to learning new user interfaces that will allow us to work with different media, thereby opening new channels for thinking and expression.

Decades ago, with the advent of television, Marshall McLuhan, in his book *The Medium is the Massage*, recognized that electronic circuitry was becoming an extension of the central nervous system. Recently, Bill Gates, the founder of Microsoft advocates using the emerging information environment as a "digital nervous system" in his book *Business @ the Speed of Thought*.

physical interface *In computer applications, that which relates to the way people work with the hardware devices.*

mental interface *In computer applications, that which relates to software procedures and thinking skills.*

spiritual interface *That which has to do with a sense of attachment, empowerment, and meaning.*

human factor *A characteristic related to people. Especially concerning how people interface with tools such as computers.*

ergonomics *The study of how energy is spent. Pertains particularly to human energy expended for doing work.*

The Look: Relating to What We See

Colored light, as opposed to reflected color, displays hues even more vividly than paint on canvas. Digital whiteboards, which use colored light, enable workgroups to interact online or in a conference room and video projection systems now also display colorful dynamic images on large screens in auditoriums.

Digital technology fills our monitors with higher-quality images received via airwaves as well as from cable and DVD. The new standards for HDTV, merging television and computing, include high resolution with square pixels, and wide aspect ratios. Wider-screen images, like those we see in wide-screen cinemas, fill our field of vision to heighten our experience. Using stereoscopic viewing devices, we can also view three-dimensional objects and become immersed in cyberspace.

multitasking *Using more than one software application at the same time. For example, some computer user interfaces permit people to work with different applications using different windows.*

Multiple windows enable us to work with multiple documents or provide multiple views of an object. We can do multitasking, and work with multimedia, capturing images with scanners or digital cameras. Thus "Multi-multi" helps us relate to what we see on a monitor, making it easier to work in cyberspace.

The Sound: Relating to What We Hear

Digital audio, coupled with good speakers or earphones, can surround us with high quality sound. We can record sound samples and modify them in ways that were never possible before. MIDI keyboards synthesize electrical sounds for mixing in many channels.

Telephony provides voice mail along with other phone capabilities such as recording verbal messages in digital files. We can play digital messages back interactively. The Internet and phone are thus merging, enhancing what each has to offer.

Voice recognition is probably the most promising user interface development. Programs, such as IBM's ViaVoice, transcribe what we say, enabling us to dictate when writing. Programs that recognize voice commands, have particular potential for applications like CADD, where we fully occupy our hands with a pointing device for digitizing. To use this interface, we train the computer to recognize our voice. (Which sometimes requires training ourselves to speak clearly and consistently.) It is becoming entirely practical to work verbally, giving verbal commands and dictation as well as receiving digital verbal responses from the computer.

The Feel: Relating to What We Touch

The keyboards that have become the standard computer user interface require typing skills, an unnatural action and posture. Pointing devices, however, provide a user interface that responds to more natural gestures. Many different types of pointing devices have emerged, such as the simple mouse, roller-ball, touch-pad, and joystick. A digitizing tablet, with a multi-keyed puck, is useful for drafting. A stylus may be pressure-sensitive to make denser lines when we press harder, or tilt the stylus, making it useful for drawing and painting. Pen computers let us mark directly upon the image on a screen. Some pen computers fit in our palms providing greater mobility that enables us to stay in touch with our physical environment.

A flying mouse, 3-D control ball, joystick, data glove, or even a body suit can make it easier to build 3-D digital spatial models and navigate in virtual reality.

The Setting

The setting is an important part of the user interface. Unsuitable surroundings can compromise human performance regardless of the interface. I find I can be remarkably more productive accessing the same programs in the privacy of my office or the study at my home in Claremont than in a noisy computer lab at Cal Poly Pomona. For me, it is even easier to write in my cabin at Big Bear Lake with a notebook computer on my lap. In fact, parts of this book have emerged on a notebook computer at my cabin or while I was in Italy.

Work Environment

In many offices people add computers and other digital devices to a tool-based work environment, using them only for data processing, word processing, imaging, or drafting. While these tools have a real-world focus and can enhance human capabilities, their potential is limited if they are used only as single-purpose tools. With a change in mind-set we can visualize ways to link information flow from these information appliances.

The desk

Using a computer workstation we can more easily exchange information between multiple applications. We can design with digital tools by developing models that address our goals and are meaningful to us. Again, with a change from a desktop mindset to a workstation mindset we can visualize workflow to develop better methods for using computer applications.

Connectivity is making it possible for yet another mind-set to emerge. Adopting a Web workstyle and Web lifestyle provides more freedom over how we use time and space. No longer do members of a design team have to work at the same time and place. We can collaborate with clients and colleagues online, exchanging information electronically. We can access shared information environments and link our operations to service bureaus that provide capabilities we could not afford to maintain on our own. Learning to share minds online enables creative individuals adopt a cyberspace mindset.

The computer

The world

Admittedly, there will always be a need for tool-based work environments where people can meet and work with tools in a physical context. Fabrication and construction require this. People also need the stimulation of real-life interaction with other people and the real world. At the same time, however, computer workstations are becoming even more useful. It is now possible to package a tremendous number of digital tools into a small box (such as a powerful notebook computer) and take this with us wherever we want to, creating opportunities to connect to a web of life that becomes an extension of our nervous system.

Technologies are merging. The voice capabilities of the telephone are becoming integrated into computers. Video and television are merging with computers. All of this is being enhanced by a rapidly growing communication infrastructure. Higher transmission speeds are making it possible to transfer a wide bandwidth of multi media in real time. Consumer electronics are mass-producing multimedia devices and Internet appliances. A growing number of people are surfing the Web participating in e-commerce, e-business, e-government and e-classrooms. Digital tools empower creative people to express themselves in multimedia and work collaboratively online.

In effect, ideas and information travel through this digital information system almost at the speed of light. What can we do with these extensions of our nervous systems? Obviously, access to tools limits our choices. However as information technology becomes less expensive, how effectively we use digital tools may have more to do with our mindset, than with accessibility.

Levels of Involvement

When we read or listen to the radio, we experience different levels of involvement. If we don't recognize the language, we get no message—we just see the letters of the words or hear only sounds. At some level we begin to comprehend patterns and learn to connect literal meaning with the message. At another level we easily comprehend the message and can relate to the nuances of what we read or hear. At still another level we lose the sense we are reading or listening. We actually visualize we are there.

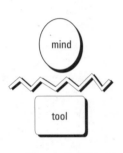

Persona involvement

In his book *No Boundary*, Ken Wilber describes different levels of consciousness. These levels of consciousness relate to what we experience when working with digital tools. At the persona level, there is a distinct separation between the tool and us. Each has boundaries. We experience this separation when we first try to use a computer application. At the ego involvement level, we begin to interrelate with a tool through its user interface. Although initially we may find ourselves not completely comfortable functioning at this level, we can at least interact with the tool. At the total organism level, a union occurs between the tool and us. Familiar repetitive tasks, like data entry, become almost automatic. And, finally, at the unity consciousness level, the tool becomes transparent. We feel as if it is part of us. The boundaries become only the limits of our awareness and imagination.

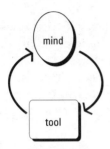

Ego involvement

We can take a familiar tool such as a pencil and readily experience these levels of involvement. For example, note how our persona relates differently to the pencil in our "wrong" hand. There seems to be a boundary between the pencil and us if we use it in an unfamiliar way. Taking the pencil in our "dominant" hand, we can involve our ego by feeling the qualities of the pencil—its length, its balance, the quality of the point, the softness of the lead. Using that pencil for a familiar task, such as writing our name, we notice the pencil becomes part of our total organism as we automatically form letters. Now, relaxing and reflecting with that favorite pencil in hand, we can let the ideas we are generating in our heads flow through the pencil. This flow may be doodles or words—whatever comes naturally from our unconscious. There is a unity between our inner space and the media space we are working with on paper.

Total organism involvement

Using digital tools (or digital toys) we can work toward deeper levels of involvement with each application. We can work past our persona to achieve ego involvement with the tool. We can bond with tools relating to them as an extension of ourselves—like something we wear, or as another being—like a "droid" or "bot." Bonding with digital tools helps us work with them creatively. By doing familiar tasks until they become routine, we begin to function with these tools as one organism. Finally, by letting our inner space merge with cyberspace we achieve unity consciousness. Be patient; this may take months—even years—of practice to achieve with more complex tools, through reasoning, vision and practice, we can expand our digital nervous systems.

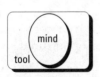

Unity conscious involvement

Here are some indications of achieving unity consciousness as we master digital tools:

- We realize that we are able to move beyond standard routines.

- We develop a sense of becoming engrossed in the application.

- We lose a sense of time during at least some of the session.

- We become oblivious to our physical surroundings during the session.

- We relate to and remember what we experience in virtual reality as if it had the richness of real-life experience.

bonding *Connecting or holding together. The formation of close interpersonal relationships involving the whole being. People can also bond with objects related to computers, and with places in reality or virtual reality.*

Object Orientation

object-oriented *Related to assemblies of information people can readily identify. People can build object-oriented models in computers. Objects can contain many attributes. Some attributes can be inherited from other objects, enabling people to build models more quickly using modules or primitives.*

object-oriented programming *An approach to computer programming that enables people to build from modules that have attributes that transfer from one program to another*

soft prototypes *Three-dimensional models built in media space using computer software.*

actual prototypes *Three-dimensional physical models.*

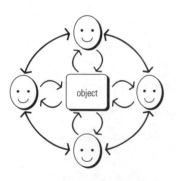

Object-oriented interaction

These encounters can become addictive—we may feel a craving to express our creative urges in this way. We should make sure we relate to computing at the unity conscious level as we would relate to normal urges for hunger or sex. We can enjoy the mental exercise of adventures in cyber space as we enjoy the exercise of adventures in reality while maintaining a balanced life.

Design professionals have traditionally worked with drawings and three-dimensional models. We now have access to digital models and maps of the realities we work with. Artists instinctively focus on the objects they are creating. Traditional art objects may be drawings, paintings, sculpture or more recently, film. Now we are seeing art emerge as constructs of information in cyberspace, creating a new genre of expression.

By creating digital models to represent what we are working on, we develop information objects. *Object orientation* provides a focus for a collaborative workgroup to share online. Objects can contain many attributes. Some attributes can be inherited from other objects, enabling teams to build models more quickly using modules or primitives. Computer programmers use *object-oriented programming* to build programs from modules that have attributes they can transfer from one program to another. We can work with objects using mathematical models, three-dimensional spatial models *(soft prototypes)*, as well as physical constructions *(actual prototypes)*.

Online object-oriented meetings enable participants to relate directly to what the group is developing together. Traditionally, design professionals have done this by using physical objects in the field, or a scale model and paper document at a conference table. Today, people accessing information objects online have another way to interact. Using digital tools, people can access shared information objects from different places and at different times and still be part of an interactive, object-oriented work session. This sets up new opportunities for groups to work together creatively. In addition to overcoming the perpetual problem of finding a common time and place to meet, digital object-oriented meetings enable participants to mark-up and transform shared objects more interactively, saving old versions as the design evolves. They also can record the transactions of information—identifying information sources as well as who changed what, when, and even why—as the interaction occurs online.

When a workgroup does meet in a conference room, they can use new media to enhance their meeting by focusing the group on the digital models they are producing and use digital tools to record their collaboration. Object-oriented meetings can make it easier to transfer the decisions made by the workgroup into products, such as reports or implementation documents, that can initiate actions.

Collaborating

Most creative endeavors today involve teamwork. Consequently, we need to be concerned about effective *collaboration*. Some people use digital tools as a way to express their ideas—to dialogue with themselves. We also can use digital tools as a vehicle for collaboration—to dialogue with others. The emphasis depends upon the nature of the creative work we do. We can focus inwardly in reflection; we can focus outwardly in collaboration. We should endeavor to do both and learn to carry on these dialogues using the digital tools that are emerging.

collaboration *The act of working together. Electronic media offer new opportunities for workgroup collaboration.*

Through collaboration we can share information, knowledge, and ideas—in effect, sharing minds. Collaboration helps us express and test designs. It also helps us organize to implement projects. We can build collaborative teams online sharing information objects in cyberspace.

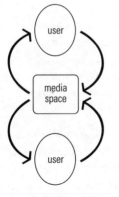

Shared media space

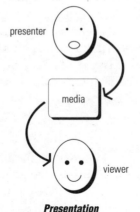

Presentation

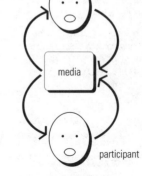

Collaboration

For example, we could work with a draft or mock-up of the actual document we are producing together. We could also work together on a financial model or a three-dimensional spatial model of whatever we are producing. Object-oriented meetings enable a workgroup to build something together in an information environment. This provides a way to link people with specialized capabilities in productive ways. Design teams may be more productive working together online, by using information objects as the focus for collaboration, especially if they can reduce the regular meetings where people just get together to talk about what they are going to do.

165

Collaborating online provides new opportunities for client participation. It can become a way to build relationships, and even engage larger audiences. The Web makes it possible to invite more stakeholders to take part and participate in projects. Engaging stakeholders online may reduce the need for elaborate presentations. More effective online interaction may improve the quality and success of a project.

Groupware

groupware Software that enables group interaction Helps coordinate large teams using computers to do complex projects.

Groupware is emerging that is designed to give online users better access to shared information. This is changing the workplace, especially in large organizations that have offices in different regions. Groupware can improve coordination and group input in making decisions. It also has potential to shorten product development cycles and delivery time.

Groupware is similar to traditional tools—such as pencil and paper, or chalk and a blackboard, or marker and whiteboard—that have traditionally enhanced group meetings. However, when working online, we do not have to be in the same place and will not run out of space as easily. Groupware provides for synchronous as well as asynchronous collaboration, so people can interact at the same time or at other times after they have had a chance to think about their contribution. Groupware can record transactions of information online or of discussions in actual meetings, providing a record of what transpires. Of course, it helps if the participants are all computer-literate when carrying on this type of interaction. But this is not essential if participant has a helper who can access the information online or if there is someone who uses the groupware to record what transpires in a meeting.

synchronous At the same time. Online communication engaging participants in interaction at the same time, as in a telephone conversation.

asynchronous At different times. Online communication engaging participants in interaction at different times, as in e-mail correspondence.

Shared Information Environments

Although currently more people have access to the mail service, the phone, and to fax—making these the preferred modes of long distance communication—this is changing. We are seeing more people use the Internet, although sensitive matters are often better expressed on the phone, or in person, to get a better sense of people's reactions. Of course, there is really no substitute for being with someone when sharing great joy, concern, or sorrow.

Local Area Networks (LANs) link computers and peripherals together in an office. There are a variety of hardware and software solutions for doing this but, to the user, the functionality is much the same. To use a LAN effectively we need to know how to navigate and access information and digital tools. It is important for each office to set up this shared information environment in understandable ways, using cognitive maps that can optimize the shared enterprise. The users of LANs should follow file naming and filing standards so others can readily share information that is also easy to backup and archive. The CD accompanying this book contains the office procedures we have at

CEDG for managing files and backing up our LAN. This simple example may provide helpful practices for managing your LAN.

The Internet, which has a standard Internet Protocol (IP), provides access to e-mail and enables us to transfer files using File Transfer Protocols (FTP). This has become the backbone for online communications for more than a decade and is growing exponentially as more people learn to use e-mail and attach files. E-mail is a wonderful addition to phone and fax. Some of the advantages of e-mail are that it is asynchronous and we can reach many people at the same time, without them being online. As with writing letters, we are able to express complete thoughts without being interrupted. Others can get the message wherever or whenever it is convenient for them. Letter writing or E-mail often elicits more focused and thoughtful responses than we may get from rushed phone conversations. E-mail also provides a record—of who expressed what and when—that is easier to save and search than letters or phone messages and logs. Having an accessible record enables us to maintain the thread of a discussion over an extended period of time, helping to more quickly resolve issues in ways that may not be possible to resolve through a simple conversation. Corresponding using e-mail can be much faster than sending letters. Once we have invested in the digital tools, the marginal cost of communicating online using e-mail is less than sending mail or using phone and fax, particularly when corresponding internationally.

We can navigate and interact on the World Wide Web using browsers such as MS Explorer or Netscape. The Web enables us to explore the links to a variety of sites with information related to our interests. We also can "drill-down" into a Web site to gain more specific information. Because the Web offers multimedia, we are able to interact not only with words, but also with graphics, images, even audio and full-motion video. With higher speed connections, this broad bandwidth of information will become even richer in content as the Web continues to grow. While initially we may have felt compelled to try to explore the entire Web, recent studies indicate that search engines may now reach less that 14 percent of its content. This leaves us with the humbling realization that the emerging information ecosystem—like the real world—has grown beyond any individual's comprehension. The challenge for each of us is to learn to navigate and use this information environment productively and also to help sustain and improve these resources for future generations. The World Wide Web is becoming an icon of the global culture. Intranets and extranets are emerging providing information environments for collaboration within and between enterprises using standard digital tools for navigating and working on the Web. Intranets were originally conceived to pro-

167

vide access only to those within an organization or firm. An extranet extends access to external parties—such as clients and consultants—through secure IP-based connections. Although the distinction between Intranets and extranets is becoming blurred, as they both take on more functionality, they each can provide rich information environments to organizations for business-to-business information transactions. Intranets are particularly well suited for project-based web sites for design collaboration and construction management. Large AEC firms such as Bechtel (www.bechtel.com), Parsons Brinkerhoff (www.pbworld.com), and Zimmer Gunsul Frasca (www.zgf.com) have begun to explore using these types of information environments for design collaboration and construction management. Internet service providers now provide extranets to smaller organizations that don't want to manage these sites themselves.

Emerging Digital Tools

Computer operating systems are converging with software for transferring files across LAN's and the Internet. Communication tools such as e-mail, Web browsers and Web authoring tools continue to evolve providing more ways to link and attach digital files.

Groupware enables people to collaborate and share information online. For example, Lotus Notes works particularly well where people in enterprises have access to the proprietary software which provides a message-centric focus that enables people to communicate, adding notes to messages and keeping track of threads of online discussions. Lotus (www.lotus.com) also provides software called LearningSpace providing both synchronous and asynchronous communication for distance learning. Microsoft NetMeeting is being integrated into MS Internet Explore to more readily provide messaging, as well as online document viewing and markup tools delivered over the Web, intranets, or extranets. Over one high-speed connection, collaborators can simultaneously conduct a phone call, a collaborative whiteboard session with redlining, and even a video conference. Microsoft (www.microsoft.com) also is offering groupware called Exchange that also provides synchronous and asynchronous communication for distance learning. Groupware that links online collaboration and learning provides design professionals with a wonderful vehicle for continuing education.

Digital white boards have considerable potential for collaboration among design professionals because much of our work is visual. This digital technology permits people in different locations to view the same document online and mark it up interactively. The document may be a page of text, a spreadsheet, a drawing, or an image. The interaction is much like people sitting down together at a table and working on the same document, except the participants no longer need to be in the same place.

Video conferencing also has potential for collaboration among design disciplines as well as with clients, consultants, and contractors. People initially think of using video conferencing for face-to-face meetings, however, adding "talking heads" to what amounts to a phone conversation may not be worthwhile, unless this involves sensitive negotiations where it is beneficial to see "eye to eye" to gauge people's reactions. Using video conferencing for inspecting a project site, or for displaying product samples, may be a better application of video conferencing. This could especially be worthwhile if it can focus collaborative discussions on the realities of a site or situation to help quickly resolve issues without the need for extensive meetings that require a lot of travel time.

As the bandwidth of online communication broadens, it is likely we will see an integration of whiteboard technology, video conferencing, and virtual reality. Already there are computer games that enable people to enter the same virtual space. Using virtual reality markup language VRML, it is possible to post digital 3-D models online. Digital tools are emerging that enable design professionals to virtually walk-through a design, navigating in real time, and interacting with others who can view and even markup the model as if it were a 3-D whiteboard. This is potentially a very exciting and beneficial way to focus collaboration among design professionals, clients, consultants, and contractors.

Design professionals have the opportunity to use these technologies internally for their projects, but also to provide online collaboration services to others. Through the effective use of online collaboration, consultants could become clients and clients could become consultants. The process of team building becomes more fluid.

Summary

Strategies for Interacting and Collaborating Online

1. **Gain access** and begin adopting a Web workstyle by learning to navigate online.

2. **Learn the interfaces** for using digital tools and expand how you relate to what you see, what you hear, and what you touch online.

3. **Find comfortable places** where you can interact productively using digital tools, giving yourself a chance to reflect and respond thoughtfully.

4. **Establish your work environments** by using stand-alone devices—such as digital cameras—in the field, computer workstations—with applications you need—in the office, as well as connecting to information environments—such as your LAN as well as Intranets, Extranets, the Internet and World Wide Web—from wherever you have access.

5. **Work towards deeper levels of involvement** to achieve unity consciousness when using the applications that are most important to you.

6. **Develop object-oriented digital models** to help online collaboration. Engage your body, mind and spirit in shaping information objects so you can enjoy the art of using digital tools creatively

7. **Build a collaborative team,** adding the specialized expertise that is needed for the project and encourage each member of the team to use digital tools creatively and develop capabilities to collaborate online if they are not doing so already.

8. **Adopt groupware** for your virtual office making sure that each member of the team has the access necessary to collaborate using these digital tools.

9. **Establish a shared information environment** for a project or virtual office, and manage it effectively making sure each member of the team can access information they are authorized to share online.

10. **Pay attention to emerging digital tools** and adopt what is useful to you and your team.

Activities

Set Up
a Virtual Office
Online

Design teams need to communicate effectively. The challenge is to correspond online in ways that enable the team to interact more quickly and easily and develop shared information objects that can be instrumental in achieving this goal.

Because computers can make us more self-sufficient, it is tempting to try to do many things ourselves using the programs we have access to. Yet, by working online, we have new opportunities to collaborate with capable people and delegate specialized tasks to specially qualified people. We can develop workgroups without the limitation of finding a common place and time to meet. In fact, your workgroup may be all over the globe. In addition, information service providers may provide capabilities not currently available in your organization.

To start working in this direction you might begin with the smallest functional workgroup in your organization (for example, this may be a project team) and develop clear approaches for working online. Use the strategies described in this chapter to help you get started. Discuss how you can collaborate effectively online. What information do you need to share? What digital tools will you use? What models can provide a focus for collaboration? You might even begin this discussion online as a pilot study for how you can interact on a real project. Once you have the issues resolved, set up a project based-shared information site for online collaboration. And once this online project site is working, see what you can do to expand this online approach to improving more of the operation of your office.

Can you see opportunities that may lead to innovations? Innovations often come from finding new sources of information (or ways of acquiring it). They can also come from finding better ways to model reality so you can deal with it more effectively. In addition, innovations may come from developing better methods for using digital tools that may enable you to improve the quality of what you design while reducing the time involved.

171

16
Manage Projects Online

By managing projects online a team can save time and money while improving quality. The quick interaction possible using information technology to coordinate large complex projects can help sustain the spontaneity and innovation of a small design studio. It also can provide important links with contractors implementing projects as well as with clients and their project managers. Forming teams that can work in cyberspace can enable creative collaboration while keeping track of progress to focus effort more effectively. Digital offices have intriguing possibilities for using information technology to manage projects. At the same time, however, there are challenges. Virtual offices need to build teams with the capabilities to network online, developing a culture that adopts web workstyles.

Project Planning and Management Methods

The classic approach to project planning and management includes: surveying the situation; defining project objectives; establishing a strategy; developing a scope of work—identifying activities and products; identifying resources; sequencing tasks and estimating the time involved; establishing deadlines and compensation necessary to carry out the project; and refining and reviewing procedures. Project management software is available to help carry out this approach. The key to successfully planning and managing a project is being able to clearly visualize the process and communicating it clearly to the project team and stakeholders.

Workflow diagrams, used widely in operations research, enable us to visualize the relationships of activities and events. We can also relate these tasks to a time line. The *CPM* (critical path method) and *PERT* (progress evaluation review techniques) use these diagrams to manage large operations. We can diagram activities as arrows (the length of which relates to the duration of the activity). By identifying events in boxes or circles we place them at the appropriate dates on a time line. Each activity or task usually relates to some event or product. For example, we can diagram the activity involved in digitizing a base map by showing an arrow on our time line which would start on the date we expect to begin digitizing and end on the date we hope to finish. The product of the digitizing task would be the map we produce, identified on the workflow diagram. By linking activities and events we can diagram sequential, parallel, and even cyclical procedures in time. This helps determine the longest (or critical) path through a project.

CPM (critical path method) *A procedure for determining which set of activities will take the longest to accomplish, hence constituting the critical path of activities in a project.*

PERT (progress evaluation review technique) *A procedure for considering what is completed on a project and relating this to a schedule.*

Strategies for developing workflow diagrams are similar to the strategies for developing information flow diagrams. In fact, information flow often parallels workflow on design projects because much of what we do is information work. One strategy is to start at the beginning of a project and diagram the path by which we think we will proceed. Another strategy is to picture the end product (or final event) and then visualize what leads up to that—working backward from the final product. Yet another strategy is to list all the activities and events involved in a project and then begin to link them, first developing subroutines and linking these modules into larger processes. As we become more familiar with various procedures, we can build these modules to shape approaches to new projects more quickly. For example, we may be able to establish some standard procedures for our discipline or team and then link them creatively when planning new projects.

Typically, project management software links graphic charts to data-bases and calendars, keeping track of our activities, events, and schedule along with the resources we have allocated. Using this software helps design our approach to a project. It helps evaluate our project to see what resources are necessary to carry it out and estimate how long it will take to do the project. In this way, we can refine our approach before we commit a real effort to an endeavor. This software can also help manage our work as we proceed. Using project manage-ment software, we can produce a variety of reports useful for moni-toring progress.

Project management life cycle

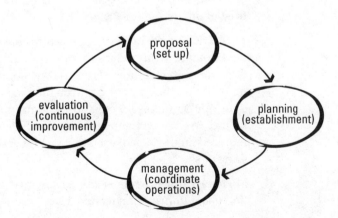

Consider a workflow diagram as simply a map or a guide through a project. Picturing alternatives is especially important if we want to make use of digital tools that are available. Often, the only way to break out of our present work pattern is to visualize alternative approaches incorporating new tools or visualizing more innovative ways to use the tools we have access to. To use digital tools effectively, we have to plan our approach. Workflow diagrams can help do that. They help to understand how to develop methods to mesh our minds with the tools we are using. Workflow diagrams also help to communicate what needs to be done so we can coordinate a team's effort. The project management life cycle involves a proposal to set up the project, a plan for carrying out the project and establishing the team, management for coordinating team operations, and evaluation of the project for continuous improvement.

Team Building

Leadership, trust, and shared minds are essential to building successful teams for online projects:

- **Leadership** involves the ability to rally support for shared goals and a vision for achieving them. To effectively build teams and manage projects, using digital tools in a virtual office, requires having more than the personal charisma of a good leader; it also requires having the communication skills to use information technology to collaborate online.

- **Trust** provides an essential foundation that cements relationships and enables people to commit to shared endeavors. Trust is traditionally derived from strong personal relationships. Carrying relationships into a virtual environment requires networking—making sure everyone is in the loop—and caring—communication that deserves the trust of others.

- **Shared minds** involve openness to consider new ideas and willingness to share ideas and information. Sharing has traditionally been done through meetings, written correspondence, and phone conversations. Today the emerging information technology provides tremendous opportunities for expanding communication and collaboration online. To effectively incorporate this new technology commitment is required on the part of each team member to learn to use digital tools so he or she can interact effectively with the rest of the team.

Michael Schrage wrote *Shared Minds*, an excellent book on collaboration using information technology.

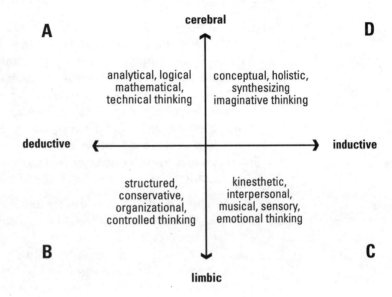

Hermann Brain Dominance model

A

D

cerebral

analytical, logical
mathematical,
technical thinking

conceptual, holistic,
synthesizing
imaginative thinking

deductive ⟵   ⟶ inductive

structured,
conservative,
organizational,
controlled thinking

kinesthetic,
interpersonal,
musical, sensory,
emotional thinking

B

C

limbic

group dynamic Interaction among
members of a team. Involves the
members' different modes of
thought.

A key challenge is the mental integration of the project team. *Group
dynamics* is an important aspect of collaborative efforts. To make the
group dynamic work, the team needs members who can contribute not
only specific expertise and skills, but also the capacity to think differently from other members. At the same time, each member needs to
understand and appreciate the value of the thinking modes of others.

Each discipline has its own knowledge base and emphasizes different
modes of thought. For example, the arts emphasize inductive thinking,
design disciplines integrate inductive and deductive thinking, while
the sciences tend to emphasize deductive thinking (although brilliant
scientists, will of course, make extensive use of insights). People gravi-
tate toward disciplines they are interested in learning about and in
which they feel comfortable with the modes of thought that are
emphasized. Excelling in any discipline typically involves a considerable knowledge base and the capacity to use different modes of
thought.

Within disciplines we use many of the same digital tools, although we may apply them differently. Specialized software applications are also emerging, tuned to the types of thinking different disciplines require.

In addition to different disciplines, there is also a range of personality types. All of this makes formulating project teams and managing the group dynamic more challenging. Certainly, these are very important considerations when meshing different minds in a multidisciplinary endeavor. Katherine Briggs and Isabel Myers developed a personality-type indicator based upon the work of Jung. They found preferences in how people energize, what people pay attention to, how people make decisions, and what lifestyles they adopt. To some extent, our characteristics may vary depending upon mood or the situation. Personality characteristics are manifestations of our feelings and thought patterns.

Given this complexity, how can we formulate multidisciplinary teams? Learning to appreciate, as well as enjoy, other people's mental capacities and personalities is key to effective collaboration, and design professionals need to build teams accordingly. It is also important to clearly articulate and agree upon roles so that each team member understands what his or her responsibilities are and can comfortably carry them out.

Modes of Operation

How we use digital tools influences how we organize and manage activities. It is important to be aware of different organizational structures, because they influence our mindset. If used well, digital tools could be liberating. Used with the wrong intent, these tools could become controlling and inhibiting.

Production-line organization

Production-line approaches to organization and management typically emphasize specialization. Each person has a position or role to carry out. Although people can make certain decisions, they are part of an organization and carry out specialized tasks. Specialization can optimize efficiency; however, taken to extremes, this type of involvement may not engage the whole mind and, consequently, can result in personnel problems. For example, absenteeism, high turnover rates, and numbing addictions are syndromes sometimes found in production-line workers. People have the potential to do much more than production-line work. Although the production line has led to a certain level of material success in the industrial age, this pattern may be changing as the information age emerges. Machines can often function better than people in production-line situations. People have the potential to carry out more creative activities that can add greater value.

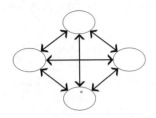

Workgroup organization

Coordinated Information Management Systems

A workgroup approach to organization and management emphasizes independent initiative. Participants relate to the whole context and the objects they are working with. This can enable people to collaborate more creatively, drawing upon more of their potential. Each person in a workgroup can assume responsibilities and contribute the unique talents he or she may have. If all participants can communicate directly with each other, this approach tends to optimize collaboration and creativity and is characteristic of research groups and design teams. The key to making a workgroup approach function effectively is communication and the right mix of motivated people. Information technology is making it possible to use this approach more widely. Workgroups can relate to digital models and other information objects online. Using digital tools creatively can stimulate cultural changes that will de-emphasize production-line mindsets, possibly leading to more humane work habits and lifestyles

Whoever our project team is and whatever modes of operation we follow, it is important to articulate the requirements for an information management system that meets our project management needs.

Robert C. Schulz, Chief of Architecture & Engineering for Capital Planning, Design and Construction for the California State University (CSU) System, envisions a Coordinated Information Management System (CIMS) to serve the needs of managing building projects for the CSU system. Through extensive studies, which were part of his architectural master's thesis at Cal Poly Pomona, he determined the following requirements:

- **Common access to project files and documents.** It essential to share information online. This can be done via file servers using FTP, web-based hyperlinks, or Lotus Notes types of document databases.

- **Database of project correspondence.** This involves monitoring transactions of information and maintaining a database to automate finding relevant data. Primavera Expedition and other workflow solutions are developing this capability.

- **Project workflow management.** The capability to manage information work can improve coordination and followthrough. An example is the request for information (RFI) and response cycle. This capability is emerging in project management software.

- **Digital signatures.** This provides a means to verify the identity of anyone entering data or transmitting information. While this is currently accomplished by issuing accounts and passwords, better techniques are emerging for establishing digital signatures.

• **Secure communications.** Information must not be changed or deleted after being entered into the project database. This must be verifiable in order for people to have confidence in the system.

• **Universal availability through the World Wide Web.** Common access available through web browsers makes a CIMS feasible. Web-based solutions, such as GCN's "Project-Specific Web Site" use this to the advantage of all the participants.

• **Offline Access to the Project Databases.** If there is limited reliable high-speed access to the Internet that insures all participants continuous access to key data needed to get their job done, it is important to enable work offline. This involves the capability to download key information, work with it locally, and then upload it again to the CIMS.

• **Support Messaging Outside the System.** All participants need to continuously get messages in a timely fashion. Internet e-mail should become the primary mode for messaging; however, linking information technology to other modes, such as voice mail/pager notification, and fax can help reach those who are not online.

Project Management Systems

Currently three key functions are emerging for online project management systems:

• **A repository for managing documents.** This online project site should be linked to both local and web-based data so that information workers have access to all the shared information and digital models they need. Software such as Arnona Software's CADviewer and ESRI's Internet Map Server are beginning to do this. Having access to documents and markup capabilities is especially important to design and planning professionals.

• **A clearinghouse for metadata and messaging.** This is necessary for handling formation flow. Software such as ProjectNet from BlueLine/Online handles these types of transactions. Messaging functions are especially important to contractors managing construction.

• **A transaction hub for managing and recording progress.** This is important for coordination and workflow. Project management software, such as the web-enabled Primavera Project Planner charts progress. Keeping a legal record is especially important to project managers and owners because of liability issues.

Business Models

There are a variety of business models for setting up online project management:

- **Apply existing digital tools** using off-the-shelf software that enables online collaboration. These can be web-based project information and messaging programs involving e-mail, pager, and fax notification, as well as real-time collaboration using phone, white-board, and video technology.

- **Purchase integrated software** with dedicated groupware functionality, such as Lotus Notes

- **Contract with a service bureau**, or rent applications such as ProjectNet from BlueLine/Online, that bundles product management tools and provides updates.

Besides determining where the data will be archived, we also need to determine who is responsible for maintaining it. Should it be on the client's server, should design professionals maintain digital information on their own computer systems as an "instrument of service"? Sometimes it makes sense to use information service providers (ISP) as third-party repositories or archive public information at institutions such as universities. Options will continue to evolve as we develop more extensive digital communities with richer information ecosystems.

digital community A group of people, with shared interests, who are linked though an online information environment.

Information ecosystem Value chains of information feeding webs of enterprises online.

Summary

Strategies for Managing Projects Online

1. **Follow the classic approach to project planning and management** that includes: surveying the situation; defining project objectives; establishing a strategy; developing a scope of work; identifying resources; sequencing tasks and estimating the time involved; establishing deadlines and necessary compensation; and refining and reviewing procedures.

2. **Diagram workflow**, visualizing how to optimize the procedures involved in the project.

3. **Build a team** cultivating the leadership, trust, and willingness to share information and collaborate online.

4. **Draw upon the mental capacity and thinking skills** provided by people from multiple disciplines needed for a project.

5. **Mesh personalities** in ways that are complementary and reinforce the team's endeavors.

6. **Develop clear strategies for modes of operation** so that each member of the team understands his or her roles and responsibilities.

7. **Articulate the team's requirements for a coordinated information** system, considering key issues such as accessing project files, recording correspondence, managing project work flow, obtaining digital signatures, securing files, providing access to data, and devising ways of supporting messaging.

8. **Determine which digital tools meet the team's requirements** to provide a repository for managing documents, a clearing house for data and messages, as well as a transaction hub for managing and recording progress.

9. **Develop a business model for setting up online project management.** This can involve applying existing tools, purchasing integrated software, or contracting with a service bureau.

10. **Refine the approach** for updating procedures as well as the digital tools used for each successive project, improving management and getting the most out of investments in information technology.

Activities

Set up a Project Management System

Online project management systems can help save time and money, while improving quality. A challenge is to transition our current practices to make better use of the information technology we have, as well as making use of emerging technology that has tremendous potential. One aspect of the challenge is cultivating the human resources. Another aspect of the challenge is identifying our needs and selecting available information technology to meet these needs in cost-effective ways.

Because each discipline involves creative thinking skills and no discipline works in isolation, most complex projects require a multidisciplinary team. For example, the construction of a building requires close collaboration between planners, architects, and engineers as well as the client, local jurisdictions, other stakeholders, and especially the construction contractor. Even projects done individually must relate to some audience, client, constituency, or other parties. Consequently, we need to be able to develop effective teams that can collaborate online and coordinate their efforts to achieve excellent results.

Information technology is emerging that offers new opportunities for managing projects online. Using the strategies outlined here we can begin to address these opportunities. We should tailor the information technology we select to our project's needs, realizing that both project needs and information technology will continue to evolve. The challenge is to carry forward what we learn from each project to guide that evolution.

181

17

Broker Digital Data

Information as an Asset

Design professionals are content providers who add value to information. As important links in the value chain, we can expand our services by using digital tools and new media to provide valuable digital data to clients and others—even after the design project is complete. Knowing how to access and disseminate information online enables us to broker digital data serving the design professions, our clients, and society.

Storing digital information requires substantial investments in hardware and software. Acquiring information represents the largest investment. Training people to effectively use digital tools and information is another significant investment. Maintaining static archives of information becomes an overhead expense; however, actively using digital information can make it an asset, creating value for an enterprise while also serving others. Finding ways to use digital information should more than justify investments in information technology.

A good place to begin is within our own practices. We can save time by using checklists, templates, symbol libraries, and stock details, as well as prototype drawings and 3D model components to automate some aspects of producing digital documents. Building upon prototype drawings or digital components provide models we can transform digitally to meet the needs of specific projects. Clever standards may improve the quality of what we produce. Saving time and improving quality translates into value. We can develop our own standards by deriving consistent information from projects to use again. We also can purchase templates and standards that are useful. For example, *MasterSpec* provides current specifications we can apply to projects. *Guidelines* provides model working drawing sheets incorporating the CSI (Construction Specification Institute) Uniform Drawing Systems and the AIA Layer Guidelines as well as dimensioning and keynote standards coordinated with checklists. (Fred Stitt makes his *Guidelines* available through the San Francisco Chapter of the AIA.)

Software applications are becoming specialized to provide digital tools, layering systems, symbol sets, and details for specific design disciplines. For example, Autodesk Architectural Desktop is specialized for architecture. LandCADD is specialized for landscape architecture. Purchasing these, or other specialized packages, provides starting points to automate routines meeting our needs.

data	hard ware & software	training
85%	10%	5%

Comparative costs (Data is by far the largest cost for GIS implementation.)

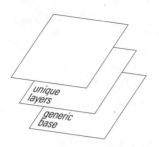

Layering Information

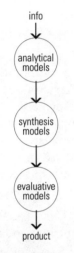

Use Models at Each Stage of the Design Process

We can save even more time, and reduce errors, by digitally transforming project information. This may involve refining a draft using word processing or a quantitative model using a spreadsheet. Similarly we can transform a digital survey, base drawing, or three-dimensional model using CADD. We can simplify drawings by turning off layers we don't need. Using multimedia software, we can transform images and even video. The challenge is to work with information creatively. Saving base information separately, so that we can always go back to the original, helps overcome inhibitions of changing it.

We can reuse information in models during each stage of the design process. Initially, our model may be useful for analysis or helping to organize and understand the information. At the synthesis stage our model can help integrate information and ideas to create new entities. As we move into the evaluative stage of the design process our model can help test ideas. Transferring information from each stage of the design process to the next can save time. Carefully naming each version of the file and saving it at each stage of the design process enables us to track this evolution. Sometimes it is useful to show the progression of thought; we may wish to go back to explore design options that we didn't develop.

Typically, design firms archive finished projects to maintain records for liability purposes, but this information may have a much greater value. Architects can use digital information to help clients set up an information infrastructure including files of buildings for facilities management. Landscape architects can provide digital files for landscape management. Graphic designers could provide digital files of logos and trademarks that the owners could use online as well as on paper. Of course, it is important to set this up in ways that protect the rights of design professionals as well as those of the clients. Some owners, and even governmental entities, recognizing the value of digital information, are requiring that design professionals make digital files available without due consideration. In the entertainment industry major studios control all rights. Digital artists, creating movie special effects, typically don't have the right to use their own work for other purposes. Working independently, and selling their work online, could enable digital artists to retain the rights to their work.

Legal Rights

The issues of copyrights, trademarks and branding are old ones. We need to protect intellectual property. It is important to get permission to use materials others have created and also credit the source properly. Copyrights pertain to quotes and written documents, as well as to data, graphics, images, sound tracks, animation, and video. While digital data is easy to copy, it also is becoming easier to obtain permission to use existing material. Online distribution and clearinghouses make stock images and sound available for multimedia productions. For example, we can purchase images, video clips, sound samples, or musical themes online. This can help compensate artists, authors, design professionals, and other information workers so they are able to disseminate more of what they produce. E-commerce is developing safe methods of making payment for these and other online transactions. E-mail is also making it easier to contact creators of digital content to ask permission to use their work.

Some information is public domain, which means you don't have to get permission to use it. This material was created either with public funds or the copyrights have lapsed. Promotional information also is typically free because it also benefits the provider. For example, manufacturers provide product information to help design professionals specify products. We can obtain this digital information either online or on CDs.

There is the ethical matter of crediting those who collaborate on projects. This becomes more complicated as we create shared information environments. Design professionals need to make sure that electronic media do not degrade creativity with rampant copying. We also need to make sure workgroups acknowledge the creative contributions of individuals. Abiding by established laws as well as moral and ethical mores can extend human values into the use of new media.

File Standards

There is a need to develop and use standards on all file types including databases, formatted text documents images, GIS maps, CADD drawings, and 3D models as well as video productions and audio recordings. Standards are emerging where there is a dominant software such as Microsoft's DOC, RTF, and XLS file formats for word processing, and spreadsheets. Autodesk's DWG file format has become a standard for CADD. Standards are also emerging where software producers have agreed upon an open architecture for digital information such as HTML files for publishing online, JPEG and TIFF for graphics, and DXF for CADD. There is a delicate balance between standards (which permit the use of data across a variety of hardware platforms and software programs) and diversity (which stimulates competition and growth of a diverse information ecosystem.)

Another approach is to develop translation software that can transfer data from one information environment to another. This would permit each entity (design professionals, clients, consultants, and contractors) to use its preferred work environment, but be able to share data. We can access Internet and intranet sites using cross-platform browsers such as Netscape and MS Explorer as well as "plug-ins" such as Java Applets and Acrobat. With these digital tools, we can view information with browsers as well as download data to work on locally using a variety of applications and then place the amended material back on the website where it can be shared.

Design and planning professionals can gather online information and work with it, adding value for clients. Essentially this is what we do for any design or planning project. There are many good sources of online information to which we can add value.

Sweet's CD and Sweet's Online provide information for products we specify on projects. We can select product information online and compile it digitally with our recommendations to clients. In this way, we can present up-to-date information with color images. Some websites have public domain collections of images useful for design communication. For example, UC Berkeley e-library (elib) has a website with images of native plants, NREL has images of renewable energy technology. Product manufacturers, such as those who produce windows or irrigation equipment, also have CDs with details that design professionals can use to correctly incorporate the manufacturer's products into construction documents.

Adding Value

Design professionals can add value embodied in products. If we design and develop a product, we can provide information online to help get the product to market. The World Wide Web is making it easier for small entrepreneurs with innovative ideas to reach niche markets all over the world. I find websites offering better bicycles intriguing. There also are some very active sites about electric vehicles that provide product information as well as enthusiastic reviews by users of these products (as there is for books at Amazon.com).

A growing amount of digital information is already available. Digital cameras and surveying instruments make it easier to gather digital information directly from the field. Digital tools such as scanners, and digitizing software also enable us to transfer information existing on paper documents into new media where we can use the information more effectively. Knowing how to quickly acquire digital information enables design professionals to save clients time and money. Our firm is doing a feasibility study for new sports complex for the University of La Verne. The intent is to explore possibilities for acquiring property and assess the feasibility for moving baseball and soccer facilities to this site. Since the university does not yet own the property, administration does not want to spend much money surveying this site.

We were able to quickly digitize available information so we can work with it in our computer system. For example, scanning a portion of the USGS Quad provided the vicinity map we needed. Scanning existing air photos provided information about natural and artificial features that served as an underlay for preliminary site planning. The air photo

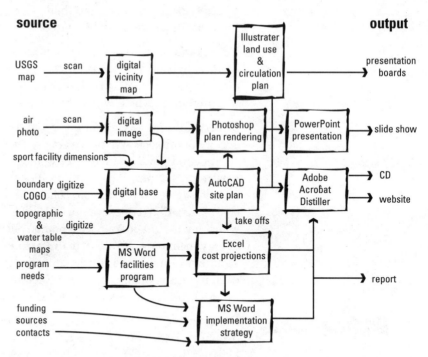

source   output

*Information flow diagram for
the University of La Verne
Sports Complex*

also became the base for a plan rendering. Studies done for previous owners of this and the surrounding property provided information about the geology, groundwater, and topography that we digitized on different layers of our CADD base. Digitizing the coordinate geometry from the legal description provided accurate property boundaries within which we could lay out the sports fields using templates we derived from NCAA league standards.

In the beginning, we worked interactively moving these templates around with the stakeholders involved, identifying concepts to satisfy their interests. We developed these concepts into a preliminary site plan to use as the basis for a feasibility assessment. The digital information we produced provided the content for presentation boards and PowerPoint presentation that would enable us to communicate the design to the University Board of Trustees and local jurisdictions. A report (and even a CD or website) could help the university gain support and funding for this project. The digital information we produced for this preliminary study has value for the design development and construction documents for this facility. Of course, we will need to add more information such as an accurate engineering survey. Digital as-built documents will have value for facility management once the project is complete.

186

Communities and businesses can link all aspects of their enterprise to an "information food chain." Beyond the local scene, a global geographic information infrastructure is emerging that offers spatial data. Designers and planners can access air photos, maps, and census information and add value for projects. We can help organizations build, or tap into, geographic information infrastructures to gain access to spatial information that is useful for decision making and management. We can bundle software and information, helping organizations set up environmental planning and management operations.

Providing Software

Programming software is another way to create value. Design offices may simply create macros to enhance certain applications. Some individuals and offices produce small programs or applets and make them available online as "shareware." This enables others to use these tools and provide some compensation to their originators. Third-party software might even be sold and added to major commercial applications.

Jack Dangermond, the founder of Environmental Systems Research Institute (ESRI), was educated as a design professional. His undergraduate degree is in landscape architecture from Cal Poly Pomona. After going to Harvard Graduate School of Design, he started up ESRI to do spatial analysis and planning. In the process, he developed geographic information systems software using polygons. ESRI now is the major producer of GIS software. Although ESRI still applies these tools, the major part of their revenue comes from developing, distributing, and supporting software. Like many software producers, they work with people who can help them enhance their applications.

Selling Information Online

Recently, I found someone selling information online that met a particular need I had. Although I have remodeled our home to be primarily solar heated and naturally cooled, I have been considering replacing our 70-year-old gravity furnace and adding backup cooling with air filtration. I read an article in the *Los Angeles Times* on the new superefficient condensing gas furnaces written by James Dulley, a professional engineer located in Ohio and was impressed by his knowledge. The article listed his website and indicated there was more information to download. At this site were well-organized "Update Bulletins." He used Adobe Acrobat to make PDF files available to download. The interesting and worthwhile bulletins contained a selection of product information along with his professional assessment. Quickly picking bulletins that addressed my needs I charged them to my credit card.

I came away from this transaction with information I was seeking as well as with a realization that here was a model for how design professionals can broker digital data. Dulley added value to this digital information by selecting and assessing HVAC systems. His site also offers a lot of other interesting and useful information for homeowners. This information is available for a fraction of what it would cost if I personally gave the same assignment to a consulting engineer. Obviously, I will still need someone to engineer the installation of the new system for my house, but I now have a much better idea of what I want to specify without having to endure the hype of a salesperson who may not be very knowledgeable. Design professionals can reach the public directly and provide valuable services.

Reaching the Public

While design professionals typically focus on business-to-business data transactions, we also can be instrumental in helping our clients reach the public, or reach the public ourselves, as more people go online. Design professionals can provide content to organizations that need to reach the public. We also can help them set up their site online or link it to websites that we may maintain as a public service or for promotion. New media enable design professionals to go way beyond an occasional press release to interface with the public. Interacting with the public online can be a wonderful way to increase the visibility of the professions. Aside from helping to market design services and provide information, it can be a wonderful way to attract talented young people to sustain the design professions, the environment, and society.

The computer and video game industry now provides more hours of entertainment, and generates more revenue, than the movie industry. Some people, educated in the design disciplines, are working in the entertainment industry creating virtual environments for computer games as well as movie special effects. There's also an opportunity for design professionals to make their digital models of environmental designs available to the entertainment industry to create settings for the virtual worlds we can experience in interactive games and see in videos and movies. Helping to improve the content could make these venues more enriching, especially for impressionable young people who become enraptured by games and entertainment.

Government organizations, such as the Environmental Protection Agency (EPA), are developing interfaces that enable the public to access geographic information on the World Wide Web. Governments are also using GIS to deal with disasters such as earthquakes or hurricanes—mobilizing activities and getting information out to the public. Businesses are using GIS to locate markets and outlets for their products, as well as to track deliveries. Cultivating customer lists is extremely important for e-commerce. Current lists constituents are extremely important to a variety of organizations. Advocacy groups, like Ecotrust, find geographic information helpful in developing a constituency to preserve wetlands and other causes. Chat rooms and distribution lists are also extremely helpful in building constituencies. There is tremendous potential to use digital data to help manage the environment and deal with anything from agricultural planning and production to managing daily traffic flow during the rush hour. Moving ideas and information digitally could even help avoid the congestion and pollution of rush hour traffic. Because of this, there are incentives for telecommuters that would also reduce energy consumption. The EPA has a web site for its Energy Star Program to promote energy efficiency in appliances, office equipment, and buildings. All these applications of digital tools have potential to make better use of limited resources while improving the quality of life.

Life Cycle of Information

life cycle *A sequence of stages that include the formation, use, and disposal of an object. Can refer to physical objects as well as to information objects developed in computers.*

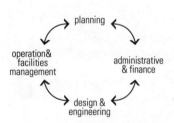

Information life cycle

Establishing information infrastructures enables people to shift from a project orientation to a transactional orientation. With a project orientation, the *life cycle* of information relates to the life of a project. When the project is finished, so is the use of the information. A transactional orientation enables people to gather information and transfer it from one endeavor to another. This gives the information a much longer life cycle, adding value that can offset costs of managing it. Information can actually improve the more that it is used, because it becomes updated and errors are corrected. Digital information can be collected as it is generated. An example of this is the way many businesses now use cash registers to compile information on sales. They no longer just ring up a single sale as if it were a separate project. Digital cash registers now collect information on a company's inventory and revenues with each transaction. This enables a company to build very accurate models of its market, inventory, and cash flow. The same could be done with spatial information. If a community could compile digital base information as changes occurred, it could build an information infrastructure that could improve planning and evaluate development proposals.

ESRI was working on major databases such as the Digital Chart of the World. The intent of this project is to put important environmental information on CDs for countries around the world to use to better address global issues. The major cost of any project is the cost of compiling information. It is possible to distribute the cost of compiling information over many projects and also share it among different disciplines. Distributing costs over many applications makes good information less expensive. Design professionals can recycle information through digital "living documents" that evolve as the world evolves. Data brokering is a growing field. There is a need for "information architects" who have an overview of many applications and know how to use information models.

Summary
Strategies for Brokering Digital Data

1. **Determine how to actively use digital information** you have acquired within your organization by making use of checklists, templates, symbols libraries, stock details, as well as prototype drawings and 3D model components.

2. **Gather digital information and add value to it.** Look for information online that you can download, or on paper that you can scan, or from the field that you can acquire digitally. Appropriately use information from other sources.

3. **Digitally transform project information,** refining it through each stage of the design process, adding to and capturing value.

4. **Extract value from information you have archived** by making it available to those whose needs it serves. Architectural information is useful for facility management. Landscape architectural information is useful for landscape management. Each design discipline can provide information to the users enhancing value and also getting feedback on a design's effectiveness.

5. **Protect legal rights** through copyrights, trademarks, and branding. Respect the rights of others by obtaining permission when using copyrighted information.

6. **Use the appropriate file standards** that make information accessible across hardware platforms and software programs. Develop clear strategies for transferring digital information throughout an enterprise.

7. **Market digital tools** you may develop to help others. Market small programs and applets as "shareware" or sell applications through commercial software developers.

8. **Provide information online** by finding audiences that value it and making it available to them. Set these transactions up using methods for payment emerging as part of e-commerce.

9. **Reach the public by helping organizations interact online** with the people they serve.

10. **Consider information cycles to distribute cost,** establishing a transactional orientation to acquire information and recycle digital information wherever appropriated.

Activities
Provide Information Online

Select information you can share with others, by matching what you have to offer with their needs. Use digital tools and new media to develop and deliver this information. Design professionals are content providers. We add value to information. We can generate value if we know how to access and author digital information online and provide it to others. The challenge is to provide this valuable information to clients and others, even after the design project is complete.

For example, the imaging and models done for a project have potential value for promotional and marketing purposes. Plans, three-dimensional models, or geographic information is valuable for facilities or land management. Online information is becoming more available and it is becoming easier to digitize existing information using scanners and other devices. Design professionals have the capabilities to develop useful models. In effect, we can become information architects helping to develop useful information. Look for opportunities to broker digital information related to the projects you are working on. Contribute intellectual capital of your culture.

18

Implement Better Ideas — Quicker

Design professionals, creatively using digital tools, can refine designs more effectively and implement ideas faster. Making designs better improves quality, thus increasing value. Implementing designs quickly saves time, thus reducing costs. Better and quicker design and development stimulates innovation and profits, providing flexibility for change and growth.

There are two basic approaches to producing better designs more quickly—design automation and design communication:

Design automation helps overcome the drudgery of repetitive tasks. Automating routines enables design professionals to use computers to simplify and speed up repetitive tasks. This has been the thrust of the initial wave of computer-aided design and drafting, but there are limits to how much can be effectively automated. The innovative nature of design does not have as many repetitive tasks as found in manufacturing or other endeavors. If used correctly, automation can free design professionals to focus on innovation and creating value through better design.

Design communication enables us to use information effectively and create virtual teams, engaging an entire enterprise. This is the thrust of the current wave of communication-aided design and data in which we use information technology to help us receive, transform, and disseminate design data quickly and effectively. This approach provides greater opportunities for advancement of the design professions, expanding and coordinating design innovation to enhance value throughout the process.

This chapter focuses on strategies design professionals can use to apply emerging digital tools and realize considerable advancement. The strategies include: connecting with clients, involving information architects, sharing information objects, using digital tools for design, distributing design responsibility, working concurrently, addressing downstream needs, maintaining virtual links with clients, using mass customization, using virtual test beds, applying quality control, extending information life cycles, and accelerating learning.

Connecting with Clients

Design professionals can connect with clients, customers, and users of the design through web-based real-time links to build client relationships and understand user needs. Bill Gates advocates "study(ing) sales data online to share insights easily." "Knowing your numbers," however, is just one of the business aspects of design. While it is important for a design firm to know where its revenue are coming from, there is much more we can do to connect with clients.

Marketing design services involves building relationships, and information technology provides more channels for doing that. Communicating online and setting up shared information sites enables us to engage whoever we work for as part of the design team, to whatever extent they wish to be, providing new avenues for gaining their input. Design professionals carry out many roles. As design or planning consultants, we serve clients; as governmental staff, we are public servants; as elected officials, or advocates, we build constituencies; as artists, we relate to patrons; as entrepreneurs, we reach markets; as educators, we serve students; as researchers, we relate to funding agencies; as writers, we reach audiences. Information technology can enhance each of these roles, providing new channels to connect with those we work for.

Information Architects

information architects People who shape our information environment by mining data, organizing information for workgroups to share, and creating information sites where people can navigate and work with data they need.

Design professionals should involve *information architects* to get the right information to the right minds at the right time. Information architects shape our information environment by mining data, organizing information for workgroups to share, and creating information sites where people can navigate and work with data they need. There is a rapid growth of information available on the web. The real need is to find ways to use this information productively while adding to it— enhancing our information environment. There are opportunities for design professionals to address this need by becoming information architects, providing these services for our own firms as well as helping others shape their information environment so they can use the web more productively. Our capabilities to create spatial experiences, coupled with our understanding of how people navigate websites, enable design professionals to make important contributions to shaping cyberspace.

Sharing Information Objects

Design professionals need to share information objects that provide critical information throughout the design and implementation process. Using shared digital information objects—draft documents, images, base maps, plans, digital 3D models, and spreadsheet models—we can work interactively with ideas and information. We can use electronic media for visualization, not just for viewing. This connectivity provides a key to creative collaboration, engaging the expertise of a project team. We can use shared digital information objects not only for online collaboration but also to enhance meetings, enabling people sitting around a conference table to work interactively using digital tools that link to information objects we can continue to share online.

Information—developed at each stage of the project—can be carried forward to the next stage and shared among the project team. Often, information objects built for one project can be transferred to another. The transfer of whole assemblies of information—whether they are spreadsheet templates, prototype sheets of drawings, or components of digital 3D models—can reduce the time and effort involved in doing a series of projects.

Digital Tools for Design

Design professionals can creatively apply generic tools as well as specialized tools to each stage of the design process. Generic tools include: word processing, spreadsheets and databases, imaging, illustration, drafting, 3D modeling, geographic information systems, and multimedia, as well as desktop and Web publishing. These generic tools are used by many disciplines. In addition, more specific tools are emerging that enable each discipline—such as architecture, landscape architecture, and product design, to name a few—to carry out their tasks more effectively, using enhancements of the generic tools bundled for their respective desktops, thus increasing speed and productivity.

Tools for each stage of the design process

Using digital tools, we can work interactively with digital models through each stage of the design process. We can capture our first insights digitally and transfer ideas and information to collaborate with others online. During the research and analysis stages we can access and compile useful digital information to analyze using computer applications. Digital tools help us synthesize and explore possibilities. We can verify designs by using evaluative tools. We can also use new media for presentations, design implementation, and marketing in ways never possible before. Using numeric controls, we can even link machine tools to computers. This enables fabrication of an object directly from a digital model without an extensive set of drawings. We need to develop clear progressions for using these tools so our effort and information will flow effectively throughout the design sequence, creating a value chain through which we provide our professional services.

Distributing Design Responsibility

Teams can distribute design responsibility throughout a value chain and delegate work by collaborating online. Information technology now makes it possible to shift design responsibility away from a central group to a distributed network that can leverage the unique resources and expertise of different team members. Benefits can include not only a reduction in time, and lower cost, but also increased quality and innovation through the creativity of close teamwork.

delegate *To entrust another person with a task.*

stop point *The point at which one task ends and another begins. Identifying stop points can make it easier to delegate tasks and transfer information.*

The strategy for transferring information is crucial. We need to learn to delegate tasks appropriately to improve productivity and job satisfaction in workgroups. Object-oriented workgroups, using share information environments that provide access to key information, can help the handoffs and enable people to work together more effectively. We need to determine stop points in documents. A simple example is to draft a document and hand it off online for someone to proofread. Then we can review the document online and refine it before we turn it over to someone who can format and print it for publication. Of course, we may want to check the final format of the document before we release it.

We can find similar stop points in all types of collaborative digital documents where consultants add their content to base information. Because digital transfers are often seamless, we may need to establish stop points by clearly identifying the tasks we are transferring and separate the work of each team member using different colors and/or layers in the digital model or saving separate files. We also need to carefully coordinate consultants' work through agreements that clearly articulate responsibilities and fairly allocate compensation.

Working Concurrently

concurrence *Happening together. Agreement; accord. Concurrence makes it possible for more people to work together with the same information.*

Design professionals can work concurrently to shorten the critical path. It used to be the case that—when creating drawings on paper—a consultant would not dare begin his or her drawings until he or she had a final base to work on. Otherwise the consultant would waste time by having to redraw everything when the base information was finalized. Today, working with communication-aided design, it is much easier to replace base layers of information, and quicker to adjust related work accordingly. Consequently, it now is possible to proceed

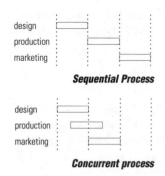

design
production
marketing

Sequential Process

design
production
marketing

Concurrent process

design chain *Enables a team of design professionals to add value ideas and information throughout the design process.*

supply chain *Addresses how a project/product moves from the contractor/producer to the end user.*

marketing channel *Addresses how a product is conceived by its maker and received by its user.*

Downstream Needs

much more concurrently. Working concurrently speeds up the process and allows more effective collaboration because each team member's work often effects others, avoiding costly changes later on.

Concurrency means agreement—getting a group to work together using common media, models, and methods to achieve common goals. Information technology provides new opportunities to do this. Consequently, work patterns are changing.

Design engineers realize that by working with computer models they can concurrently develop the design and production processes. This can reduce the time it takes to go from research and development to production. It also can result in better design. Usually engineers will use quantitative models for the initial research and design development. Once they have generated a form, they can model it three-dimensionally, using a computer. This provides an object that the design team as well as the production and even the marketing team can relate to and work with. This object continues to evolve until it becomes a reality. Using computer models, engineers can make "soft prototypes" –digital models they can easily change. Using numeric controls, they can make "hard prototypes"—physical mock-ups of objects to see how they fit together. Then they can set up manufacturing processes with greater confidence.

Concurrency also has potential for construction. Design/build and fast-track construction are nothing new. But information technology is improving the way we can use these approaches to building.

While this book focuses on the *design chain* through which design professionals add valuable ideas and information that become part of their services/products, there is also the *supply chain* which addresses how the project/product moves from the contractor/producer to the end user. To mesh with the entire enterprise, design professionals need to look at the downstream needs of the contractor/producer and end user. Designs often are compromised by the contractor/producer supplying materials he or she finds available at the time. Information technology makes it easier to communicate directly with the manufacturer, supplier, contractors, and producers to make sure each component is properly specified and incorporated into the finished project/product and also available when needed.

To do this effectively takes a different mindset—one focused on what Louis W. Stern, professor at Northwestern University's Kellogg Graduate School of Management calls the *Marketing Channel*—which addresses how a product is conceived by its maker and received by its user.

Virtual Links

Design professionals can maintain virtual links with clients and customers. We can now create web sites less expensively than producing and distributing newsletters or catalogues while also providing more opportunity for interaction. Creating a presence online can help build relationships and a sense of community. This can help design professionals acquire a better understanding of customer needs.

Aside from sharing interests, these virtual links become the network that connects virtual teams accessing and sharing information to collaborate online. This enables us to bring together greater expertise more quickly to address a design project. Large firms no longer maintain an overwhelming advantage by having expertise in-house. Small firms can draw upon experts throughout the world using information technology.

Mass Customization

Design professionals can use mass customization to meet the needs of specific users while still differentiating brands. Our design or marketing/delivery systems, key components, or the capability to modify the designs for specific users may be qualities that distinguish, or "brand," what we do.

As Edward O. Welles, a senior writer for *Inc.,* points out, the initial wave of e-commerce successes dealt with commodity-based transactions. Websites help people find the best price for goods that are widely distributed. An example of this is the purchase of computers, which have become a commodity. Computer components—and their specifications—are often more important than the brand names of the system, enabling people to customize the computer system to meet their needs.

The second wave of e-commerce has to do with specialization and adding value that enables buyers on the net to know more about what they are purchasing than ever before. This provides opportunities for design professionals because we can provide specialized services and products, as well as information content that people need. Until the advent of the Internet, it had been cost prohibitive to provide custom design services to the mass market. Amazon.com and Garden.com are two very interesting sites that represent organizations that are successfully catching the second wave of e-commerce, providing more mass customization to meet customer wants and needs.

Amazon.com does not produce books or even have a physical store, yet it provides a service that is revolutionizing how books are being sold, and they have developed a presence that enables many people to quickly and easily find them online. After we log on to their website and register, we are greeted by name. Each time we search for a title of a book, we are offered suggestions related to our interests. Once our billing information is entered, we can order with one click of the mouse. And they also update us on the progress of filling our orders.

Garden.com addresses gardening—America's most popular hobby—a $46.8 billion dollar industry that is more than twice the size of the book industry. Jamie O'Neill, one of the founders of Garden.com, sees it as "a distribution channel based around the customer's needs." They have been able to source quality products and get them to customers quickly and reliably. In doing so, they are building a brand name for their website to guarantee repeat business by establishing a community through the information content they provide. What is particularly amazing about this site is that they are able to pull this community together in a highly segmented industry linked to regional and local phenomena. Using language and imagery, this personalized website—where they sell over 16,000 different items—attracts the average visitor to linger at the site for a half an hour at a time, according to Kelly Mooney an Internet retail consultant. An information-rich site that establishes a strong brand—with good proprietary products that meet customer's needs—can do well on the web. Working closely with Federal Express, they have also developed a highly automated distribution system that ships orders directly from producers to customers—without the need to warehouse perishable items such as plants—while allowing the customer to track their own shipments.

Virtual Test Beds

Design professionals can use virtual test beds involving evaluative models in cyber labs or virtual environments that can test new designs and address changing user needs in real time. This could be a 3D model simulating sight lines, solar angles or daylighting, as well as many other phenomena that become important design determinants. It could also be spreadsheet modeling of cost, energy, or water budgets. There are many other useful digital models as well. By designing with more information we can use less material and energy.

A distinct advantage of working with digital tools is the ability to quickly explore alternatives and test them using the emerging array of evaluative models. We can optimize what we create in virtual reality, making it less necessary to produce physical prototypes. In some situations, where it is not possible to build a prototype, evaluative tools provide ways of assessing impacts before we proceed with developmental actions. Good evaluative tools can reduce the cost of research and development by shortening the time it takes to get new ideas to an audience or to market. If used cleverly, evaluative models can more thoroughly test design ideas. By going through iterations—refining the design digitally—what we deliver is not always a prototype, but a more tested design.

Quality Control

Design professionals can address quality throughout the design and implementation process. Deming and others identify quality as that which satisfies or exceeds the customer's or clients' needs and expectations. Quality assurance involves a planned pattern of all the actions necessary to assure a project (or product) will conform to established requirements. Quality control implements the quality assurance plan. Total quality management ensures that all components of the quality assurance programs are being implemented throughout the enterprise.

A survey done by Robert C. Schulz, Chief of Architecture & Engineering for Capital Planning, Design and Construction for the California State University (CSU) System, found that architectural firms primarily use quality control to improve client services, minimize professional liability, and improve office productivity. The survey identified other benefits such as reducing the conflicts and increasing the quality of construction documents to minimize change orders and delays, as well as improving the knowledge and skills of staff. Information technology can be particularly helpful in coordinating the team so that the work of each discipline fits together more successfully. It can also help establish requirements, coordinate drawings and specifications, and track changes.

The architecture, engineering, and construction industry (AEC) has established some tools for quality assurance. These include published quality assurance guidelines such as those available from Redicheck and Fred Stitt's Guidelines. The International Organization for Standardization develops and publishes ISO 9000, addressing how an organization manages its efforts to achieve quality. These manufacturing, trade and communications standards have become the core of the international efforts to maintain quality levels while doing business globally.

Design automation tools can assist the development and coordination of documents. These include using computer-aided design to establish prototype reference drawings or backgrounds. Condoc and other key noting systems coordinate drawings and specifications. MasterSpec provides updates for improving specifications. Many professional firms and design teams establish their own automated systems for handling standard drawings, details, and specifications, as well as internal procedural guides for automation.

Other important mechanisms for quality insurance included a project log for material and equipment selections as well as a log identifying design issues and their resolution. In addition, it is important for design professionals to maintain a database of instances and causes of errors and omissions. Some professional liability insurance companies provide training programs for design principals, drawing upon shared experience to help reduce errors and omission claims.

Extending Information Life Cycle

Design professionals can extend information life cycles by coordinating and sharing information related to programming/product definition, production, and operations. Information is the most expensive component of design systems. Learning to recycle information can amortize its cost over more projects. Information technology enables us to increase the value that can be derived from information.

As already described in Chapter 17, design professionals can add to the information infrastructure that is useful for land and facility management. We can develop templates, prototype drawings, symbols and 3D components, as well as cost information and lists of suppliers that can greatly reduce the time involved in subsequent projects. By cleverly drawing upon procedural information, we can recycle the success of each project.

Accelerating Learning

Design professionals can accelerate learning through constant feedback. In this way we can learn from our mistakes and correct them quickly. Connectivity can provide constant feedback from clients, customers—the users of the design—stakeholders, and contractors. We can engage them in real-time interaction and develop the flexibility to adapt and respond, improving the design as we proceed.

Traditionally, learning occurs in several contexts—in professional schools, in design offices, and also throughout entire enterprises. In professional schools there is a tendency to emphasize research and analysis of preliminary designs—the initial stages of the design process. In professional offices there is a tendency to emphasize production of construction documents—which represent a major portion of the work that needs to be done. Now, information technology is making it easier for both students and design professionals to expand their view to entire enterprises and develop more balanced approaches for using feedback from evaluative models and user response to improve designs. More effective use of feedback would enable us to more quickly improve the quality of what we design.

Most important, learning helps build teams. We can use what we learn—not only the knowledge base of our discipline, but also what we learn about better methods and strategies—to build the intellectual capital of organizations we work for. Communication skills, using information technology, have tremendous potential, but they still rely on the human skills of working together. We can use what we learn about working together to build relationships that can evolve throughout our professional careers.

Summary
Strategies for Implementing Better Ideas — Quicker

1. Connect with clients, customers, and users of the design through web-based real-time links to build client relationships and understand user needs.

2. Involve information architects to get the right information to the right minds at the right time.

3. Share information objects, providing critical information design professionals need, throughout the design and implementation process.

4. Use digital tools for design by creatively applying generic tools as well as specialized tools to each stage of the design process.

5. Distribute design responsibility through a value chain, delegating work by collaborating online.

6. Work concurrently to shorten the critical path and promote collaboration.

7. Address downstream needs in the design and supply chains.

8. Maintain virtual links with clients and customers.

9. Use mass customization to meet the needs of specific users while still differentiating brands.

10. Use virtual test beds involving evaluative models in cyber labs or virtual environments that can test new designs and address changing user needs in real time.

11. Address quality throughout the design and implementation process.

12. Extend information life cycles by coordinating and sharing information related to programming/product definition, production, and operations.

13. Accelerate learning through constant feedback.

Activities

Create Better Designs — Quicker

Design professionals continually produce original work, providing innovation that enhances the value of each project. The challenge is to creatively use digital tools to be able to do this better and quicker, so that we can improve the design professions while remaining competitive.

We need to look for ways to automate, simplifying repetitive steps. Even more important, we need to look for better ways to communicate, coordinating entire enterprises. With the emergence of the World Wide Web there are new possibilities for productivity and the realization of a project's greatest potential using information technology, not only throughout the design and supply chains, but also throughout the entire market channel

Reducing the time that it takes to complete a better project, or get a better product to market, provides distinct competitive advantages. Quicker/better approaches are keys to improving the design professions and realizing returns on our investments in information technology. Continuous improvement is a never-ending process. Learning may involve ascending steep learning curves to reach higher levels of performance, and from these levels we gain a better view of still higher levels.

19

Create Virtual Realities

Visualization Needs and Opportunities

Design professionals, creating virtual realities, can enable people to experience physical designs as well as other phenomena. Three-dimensional modeling avoids having to translate 3D form into 2D drawings and back again to experience or build 3D designs. Virtual reality, involving 3D form, also enables people to explore places that may (or may not) exist. Virtual environments are emerging on the Web and CDs that we can view on monitors, through interactive simulators, in computer games, and at theaters or on DVD. For example, through 3D imagery gathered by a lunar landing module, we can explore the moon. We can immerse ourselves in simulators and interactive games using 3D viewing and navigating devices to find our way in real time. We can view movies with special effects and digital imagery so convincing that we suspend our disbelief, thus experiencing virtual reality.

Relatively few people easily visualize 3D phenomena when viewing a 2D set of plans. (As design professionals we spend years refining this capability.) Enabling people to explore 3D models helps them visualize what a construct will be.

As design professionals we typically do redundant orthographic drawings—plans, sections, elevations—some of which cannot accurately describe the complex forms found in nature, or the compound curves emerging in contemporary product design and architecture. The most common errors and omissions have to do with 3D spatial conflicts—for example, beams interfering with HVAC ducts—or conflicts between information shown in plan and in section. Digital 3D models could help avoid this. Orthographic drawings also tend to result in boxy designs, simply because flat building pads and boxes are easier to draw and to build from 2D drawings. Three-dimensional models can describe the curved forms we find in landscapes or in efficient curved structures such as arches, vaults, domes, tents, bowls, and other complex building forms that could better fit the land. Three-dimensional models are especially useful for designing spacecraft, airplanes, boats, trucks, buses, automobiles, bicycles, computers and other curved product designs. Using 3D models can help us understand how complex forms fit together and generate more accurate area and volumetric take-offs.

CAD systems using 3D have outperformed systems using 2D in "CAD Shootouts" demonstrating that it can be quicker and easier to model forms than to draft them. Digital tools for working with 3D are undergoing amazing development. In the past ten years there have been significant improvements to the user interfaces and the features available, making it easier to create, modify, and display digital 3D models than before. Faster, cheaper computers now make digital 3D modeling affordable.

Designing with digital 3D can be fun. It enables design professionals to explore space and form in a more direct manner, drawing upon more of our intuition, as we interactively try out new possibilities. It is becoming easier to take apart and re-assemble digital models than carve up cardboard models or draw endless perspectives, and is especially easier and less expensive than building actual prototypes. And we also can interactively view digital 3D models. At a conference of the International Design Communication Association, Yasser Maliaka, who was an architectural student in Carl Rald's multimedia class at the University of Arizona, demonstrated how to import a digital architectural model into a popular game program and navigate through it in real-time. The possibilities are astounding.

Trends and Applications

There is a steady movement from 2D to 3D technology in engineering disciplines involved in product design, as well as in the architectural disciplines, and in animation. "Simply put, it's just a better way to design," writes Jon K. Hirschtick, CEO of Solid Works Corp., who sees the following trends in engineering as we move into the new Millennium and a new era of design: Affordable solid modeling is leveling the playing field between large and small design teams; better digital design tools are enabling engineers (as well as other design professionals) to do more without drafters; and the need for traditional drawings is declining. Solid modeling facilitates design communication. Better visualization results in fewer design revisions and speeds delivery of accurate designs and design quotes from others throughout the supply chain, saving time and money. Curved forms can fit better and have aesthetic appeal, making them more salable. The ease of use inherent in today's solid modeling tools enables designers to spend more time designing.

There are many applications for 3D models providing benefits that can help amortize their cost. The most important is that solid models provide top assemblies—the shared information objects—that collaborative design teams can relate to online. Architectural firms find that multimedia presentations of 3D visualization can help obtain design commissions as NBBJ did with the Staples Center in Los Angeles. Virtual reality can be helpful in gaining approvals during the review process. It can also be used to evaluate products under development or for marketing, promoting facilities or products before they are complete. And after completion of a project, digital 3D models can be useful in information kiosks, and for facilities management, or to provide virtual walking tours people can access on the web promoting places such as resorts, sports facilities, or colleges. In addition, virtual reality can be useful for education, exploring phenomena. Virtual representations of historical artifacts help build awareness. The University of Missouri is developing a 3D virtual model in preparation of a 2004 Cyberfair, commemorating the 1904 St. Louis World's Fair. Virtual reality is useful for training in simulators, teaching people to fly airplanes or pilot large ships. Simulations of events are also used in legal cases to show juries what happened in complex situations. Also, virtual reality is becoming a place for people to play as more and more computer games and movies make use of these information environments.

Paradigm Shift

We are seeing a paradigm shift as we move into the digital age. Over hundreds of years, the design professions have refined the use of two-dimensional orthographic drawings produced and reproduced on paper (and other flat media) that have to be physically transported. Now that is changing. Today, we can work digitally in three-dimensions and transmit 3D models online. To do this requires a fundamental shift in the way we do our work.

Traditionally, specifications, along with 2D working drawings, have been the primary implementation documents. Related to these plans sections and elevations, are keynotes, schedules, details, sketches, and other submittals. Plan checks, fabrication, and construction are all geared to these documents and require reproducing volumes of drawings on paper that must be transported and stored at considerable expense.

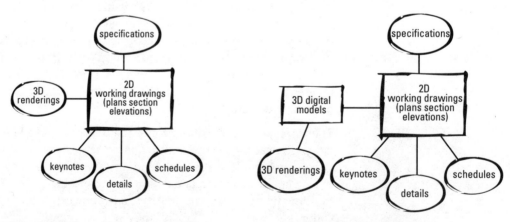

Traditional organization for implementation documents

Transitional organization for implementation documents

Today, some design professionals are switching to digital 3D models. Currently, the models tend to supplement traditional implementation documents. The 3D model may only be used for presentations of the design, or as a navigation aid to help people find their way around the maize of plans, sections, elevations and details. Using both 3D and 2D often results in considerable redundant effort.

In the future, it should be not only possible, but also preferable, to design and implement many projects using 3D models. We can incorporate specifications along with digital 3D "constructs" or "top assemblies" as the primary implementation documents, and coordinate them with keynotes, schedules, and 2D clarification drawings showing details not covered by the 3D model. We could even add verbal annotations and photo images linked to components, as well as animations showing how the pieces can be put together. We could submit CDs or DVDs to plan checkers or provide them access to these models online, so they could mark them up digitally, even using parametric procedures to expedite "model checking." Fabricators and contractors, who are part of the supply chain, could also be provided these digital documents so they could do their take-offs using digital tools that enable them to extract the information they need more accurately than they could from paper drawings. The same digital documents could also be used in the shop or field and accessed through a computer at the job site where they could be updated as part of the job record and used to provide as-built documents for management of land, facilities, or products. The digital model could be used with numeric-controlled machines, or portions of the documents could be printed out for use in the field or on the shop floor.

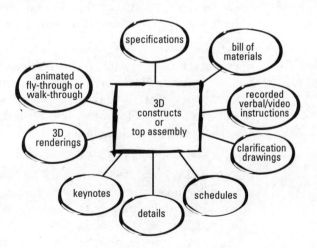

New paradigm for organizing implementation documents

207

To accomplish this paradigm shift will take the cooperation of entire enterprises, but for many types of projects there are competitive incentives to drive this change. There is considerable potential to improve both speed and quality of what we design and produce. In addition to making use of the digital tools that are emerging, to shift to a 3D paradigm will require establishing new standards. These involve not only file standards, but also standards for layers and 3D constructs, including XML, to extend links on the web for business-to-business and e-commerce development. The AEC industry needs to develop standards for referencing 3D digital models building upon what has already been established by the CSI, AIA, and other organizations for 2D digital drawings. It will be necessary to train extended teams, including design professionals, client representatives, plan checkers, contractors, fabricators, inspectors, and others—even in the marketing channel—providing them access to 3D digital documents and helping them learn to use 3D digital models for their purposes. Initially, it may be easier to transport larger digital files on CD or DVD. As the bandwidth of high-speed communication increases, and strategies for data compression and real-time transfers improve, we will be able to share more of the design and implementation documents online, avoiding the need to physically transport and store redundant copies of files or paper drawings. Other industries, such as education and entertainment, could readily plug in to the 3D digital models of the virtual and real worlds we can create and derive even more value.

3D Digital Modeling Programs and Tools

Programs for working with digital 3D models are evolving in different ways. Some programs are 2D with 3D capabilities, others are 3D with 2D capabilities, still others are dedicated modeling programs that produce virtual models including attributes. There are programs for visualizing 3D landforms. In addition, there are dedicated 3D modeling programs and 3D animation programs as well.

Autodesk's AutoCAD is an example of a program that began with 2D tools and has developed 3D capabilities, although digitizing is primarily done using 2D tools. AutoCAD can transfer files to 3D Studio, another program by Autodesk that does rendering. Microstation's Triforma is now a 3D program closely linked to 2D drafting that also includes rendering capabilities.

An example of a dedicated modeling program is Graphisoft's ArchiCAD for architects, which uses a "virtual building" as a top assembly linked to plans, sections, and elevations, as well as to a bill of materials keeping track of key attributes of the components of the building. ArchiCAD also provides some rendering capabilities. Dedicated modeling programs, such as Mechanical Desktop 3D Parametric Modeler by Autodesk, are emerging for mechanical engineers doing product design. ESRI's Arcview 3D Analyst can drape air photos over digital landforms and integrate other 3D models, enabling the user to visualize how the built environment relates to the terrain.

Form Z by auto-des-sys, is an example of a dedicated 3D modeler that can use files from 2D programs, such as AutoCAD, and generate sections along with other views that can be rendered. Form Z is being used to model 3D forms ranging from natural objects to buildings to product design, as well as 3D forms in movies such as *Star Wars Episode I: The Phantom Menace.*

In addition, 3D animation programs are emerging in the entertainment industry. For example, Avid Softimage does 3D modeling, along with animation and rendering, with raytracing and radiosity, creating the moving images and some of the special effects we see in theaters and on television.

There are tools for digitizing, manipulating, and viewing 3D wireframe, surface, and solid models. Each program has its own blend of tools for the functions it provides. For example, inexpensive programs—such as Rhinoceros—using NURBS (Non-Uniform Rational B-Splines) accurately and flexibly describe any complex 3D organic free-form surface or solid. Boolean operations permit the union, difference (subtraction), or intersection of 3D objects, making it easier to construct digital forms.

Using photogrammetry to derive 3D digital information from air photos makes it possible to quickly digitize landforms. 3D scanners are also emerging that digitize existing 3D objects. We can build new objects by using key parameters to generate form by simply entering numbers into a form generator for standard elements, like stairs, or by manually digitizing in 2D and then extruding or rotating the form to make it three-dimensional. We also can build with 3D forms by modifying or putting together primitives such as circles, cubes, cylinders, pyramids, and cones. Another strategy involves building with a set of components—such as walls, windows, and doors—ready made for architectural applications, or 3D trees and shrubs for landscape architectural applications. Cybermeisters in design firms, as well as students in professional design schools, are becoming rather adept at building anything in virtual reality.

Many analytical and evaluative tools are emerging for use with digital 3D models. Some programs link components of the digital 3D model to databases evaluating attributes such as materials and costs. The digital 3D model is useful for simulating what is seen from any location or visualizing how components move through their full range of motion. Simulating the movement of the sun helps to better understand solar orientation. Programs, such as Accurender and Autodesk Lightscape, simulate luminosity from daylighting and artifical lighting. Finite element analysis tools enable engineers to size structural members. Programs that simulate aerodynamics help optimize the designs of airplanes and automobiles, while simulating fluid dynamics helps improve the design of high-performance boats. These and other programs enable specialists to optimize specific determinants of forms.

Navigating tools are much easier to use than generating tools since they involve mostly viewing. Some programs make it possible to view the model with multiple viewports so that one view can become a reference for navigation while examining closer detail in another viewport. It is sometimes helpful to look at 3D and 2D views simultaneously to find what we want. With relatively little experience, members of a design team can learn to navigate and markup digital 3D models, extracting information they need to work with in their own applications.

Most programs can be linked with rendering tools that generate a more realistic appearance and create walk-throughs or fly-throughs, providing spatial experiences that can be transferred to video, CD, DVD, or even the web. Stereo lithography and laser cutters can quickly produce physical models from digital 3D models. It is also becoming possible to invite users into virtual reality using stereo viewing goggles and navigating devices such as data gloves. It is even possible to project holograms, enabling people to view 3D images derived from beams of light. In these ways almost anyone can view 3D models generated digitally.

Through the Looking Glass

"In another moment Alice was through the glass, and had jumped lightly down into the Looking-glass room."
Lewis Carrol, *Through the Looking Glass*, 1872.

We visualize fantasies by dreaming, or through meditation. We also can fantasize by responding to external stimuli, such as a sunset, a movie, or a musical performance. Now, using virtual reality, we can actively explore new realities. Through a looking glass, such as a monitor, we can visualize and interact with virtual realities.

We enter virtual reality by suspending our disbelief. Many digital tools are emerging that can help us do that. One approach is to experience virtual reality by using a simulator that has realistic controls. For example, a flight simulator for training pilots interactively links controls to a virtual situation as it would appear through the windshield of an airplane. We also can interact with virtual objects using 3D software. For example, a 3D web site marketing cell phones could include a digital model of the phone with which people could interact. Prospective buyers pushing buttons on the virtual phone could explore its features on the screen, enticing them to order the phone online. Or repairmen could see how an

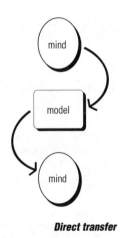

Direct transfer

mandala *A video setting you walk into and interact with.*

avatar *a virtual being that can cross over into cyberspace.*

object they need to repair is assembled and quickly order the correct part by simply clicking on it. Design professionals could even invite clients to experience a 3D spatial model of a building, urban space, or garden and get feedback from the client or users before the environment is built in reality. Relating to avatars by using real-time interactive animation (even controlled by a body suit) can help us navigate in virtual reality. In effect the avatar can become our being in cyberspace. Game software is making it possible for several players to simultaneously be in a cyberspace environment, even over the web. Yet another approach is to use a mandala to interact with a video setting that we walk into and interact with. We become part of the image on the TV screen. Weather forecasters on TV use this technique to walk through video images of weather maps. This also has potential for interactive video. Using these approaches we can enter virtual reality and even interact with other people there, creating shared experiences that have tremendous potential for training, entertainment, e-commerce as well as online design collaboration.

By moving back and forth between virtual reality and reality, we can verify the information we are working with, testing the ideas we are generating. Moving back and forth is also necessary to give more meaning to what we visualize in virtual reality. Experiencing the actual reality of a situation enables us to attach our experiences to the model. Then we can go into virtual reality, related to that situation, and imagine a richer experience. For example, if we have hiked the terrain, we can get much more out of a terrain model. This richness of experience is important to creativity and communication. Virtual reality is not entirely in a computer. We visualize it in our imagination.

Visualization integrates our body, mind, and spirit. In reality, we use visualization to relate to our surroundings, consider possibilities, and coordinate actions. In virtual reality, we can use visualization to relate to cyberspace. We can extend our creative energy to interact with objects in the new worlds we can model.

Summary
Strategies for Using Virtual Reality

1. Select digital tools that suit your discipline and began to set up a virtual reality model shop using computer(s) with sufficient speed and capacity to do the digital 3D modeling.

2. Developed a team that includes model builders capable of generating the design forms as well as design professionals capable of navigating and optimizing the forms by making real-time refinements.

3. Acquire digital 3D information such as terrain models (available online or from photogammetry or from digital surveys), scanned 3D objects, 3D symbols, clip art and other stock components, and integrate this with your own forms as you begin modeling your designs.

4. Develop techniques for sharing 3D digital mass models of preliminary designs with clients, stakeholders, review bodies and other entities, such as the prospective market or users and incorporate their feedback. This could include presentations of walk-throughs or fly-throughs and other ways of experiencing the spaces or form on video, on CD, or online.

5. Cultivate consultants capable of working with digital 3D models and begin to use 3D digital design models as a focus for collaboration, adding value through design development. Use analytical and evaluative models to optimize the design. Integrate photo images, exploring colors and textures of materials. Continuously refine the virtual prototype.

6. Develop a "construct" or "top assembly" as the primary implementation document and coordinate this with keynotes, schedules, clarification drawings and details, not shown in the 3D model—but necessary for implementation. Gain feedback from those down the supply chain to make sure that they have what is needed to effectively implement the project on schedule and within budget constraints.

7. Help plan checkers and regulatory agencies use 3D digital models to verify that your project meets their requirements. Provide supplementary documentation and corrections where necessary.

8. Work with contractors and fabricators capable of using the 3D models to implement the project. Establish a project management system based on effectively using the 3D digital models along with other contract documentation.

9. Share the virtual reality with those interested in the reality. Use digital 3D models to help market the product or promote the facility. Incorporate client/market/user feedback and re-use the models as a starting point for further testing and refining the design. Apply the models to land/facility/product management using the virtual reality of what the design "could be" to help manage the reality of "what it is."

Activities

Produce Virtual Realities

Design professionals create virtual realities that can be turned into reality. The challenge is to use this capability to explore designs more completely before they are built. We can do this by setting up design and supply chains that feed on the virtual reality of digital 3D models. Working with virtual prototypes, we can gain useful feedback from the market channel, helping us to continuously improve designs before they are built.

Use the strategy in this section to help realize the potential of working with virtual prototypes. Virtual reality could provide more effective feedback for continuous improvement throughout the design and supply chain. Virtual prototypes used in the market channel can help further refine the design. For example, virtual prototypes can provide digital data needed for numeric-controlled tools to fabricate components, facilitating mass customization and reducing the time that it takes to develop a design and get it to market.

Design professionals also can create experiences people could explore in an information environment. These can be educational experiences, as well as digital games, or other entertainment. Design professionals using digital tools could be information architects shaping cyberspace in new ways for people to explore. The challenge is to build new—or reconstruct old—places in virtual reality for people to experience electronically. This provides ways to explore alternative futures, or play games that transcend known realities. This also helps people examine and understand the past—deriving meaning that can enhance our culture.

20

Use Multimedia

Multisensory Phenomena

Multimedia Applications

Using multimedia, design professionals can communicate more facets of a design in ways that audiences more readily comprehend. We can go beyond presenting "pretty pictures" by providing stimulating experiences using the digital multimedia we have available.

Phenomena we experience are multisensory. Digitally capturing images—flat work using scanners, still images using digital cameras, or full-motion pictures using digital video—we can record what we see. Digitally recording sounds—voices through voice mail or using sound cards with voice recognition software; or music and ambient sound samples using digital recorders—we can record what we hear.

Using digital tools, we can work creatively with the multisensory phenomena we capture. In reality, all experience is multisensory; in virtual reality experiences are derived from multimedia. Relating to what we see and hear provides greater impact, more comprehension, and better recall than relating to what we may only read.

Multimedia, like telecommunications, does not occupy a separate niche, but is integrated into many applications. "Multimedia software," such as Adobe Premiere, provides tools to integrate images, animation, videos, and audio. However, even word processors, such as MS Word, include a simple graphics editor. Desktop publishing programs, such as QuarkXPress or Adobe Pagemaker, allow even more elaborate formatting for integrating text and graphic files from many different sources. With e-mail, such as Eudora, we can attach graphic files as well as voice messages. Imaging programs, such as Adobe Photoshop, enable us to manipulate photos and integrate text. Graphic programs, such as Adobe Illustrator, handle text and line graphics, enabling us to lay out posters. CADD programs, such as AutoCAD, provide both 2D drafting and 3D modeling capabilities. Many CADD programs and GIS now enable us to integrate images with plans and maps. 3D animation programs include rendering capabilities. Nonlinear video editing integrates video and sound in new ways. Music programs enable us to mix sound. All of these evolving applications with open multimedia channels for expression we can work with throughout the design process.

Integrating multimedia into the design process makes it easier to use for presentation and interaction. We can draw upon what we compile digitally and package it in ways that are appropriate for each audience.

214

Multimedia Presentations

Making good presentations requires more that just showing what we have been working on. We need to carefully select what is relevant to the topic and of interest to a particular audience. There also needs to be a story structure that engages the audience, creating an experience that brings them from a beginning point or situation, through some key points or climax, to an endpoint or resolution. Without something of a classic story structure, we may be presenting a string of information that goes nowhere and has little meaning. The challenge is to make presentations that get beyond being just focused on showing what we have been doing. "First I got a basemap and then I..." Instead, design professionals should assume a role through which we "set a scene" and clearly make compelling and relevant presentations, engaging audiences in ways that enable them to derive meaning from experiencing the presentation.

Story structure is important for writing as well as for making good multimedia presentations. I have endeavored to structure the parts of this book beginning with methods addressing "what to do" and then building to case studies addressing "how to do it" and ending with strategies for "realizing the potential." Each chapter begins by identifying needs and opportunities; building to a presentation of methods, approaches, or strategies; and ending with guidelines and activities the readers can use in their own practice to derive more meaning and value from what is presented.

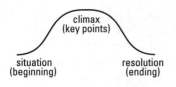

Story structure

With multimedia presentations, story structure is of utmost importance. Our use of multimedia should not call attention to itself in ways that are distracting. Even though emerging technology provides digital presentation tools, such as MS Powerpoint or Adobe Premier, that enable us to use new media in many ways, we should keep presentations simple and effective, striving to tell the story in the clearest and most compelling way. We become the writer and director of a small production. In the final assessment most mature audiences will seek meaning rather than just momentary experiences involving more "bells, whistles, and explosions with flashes of light." By using multimedia presentations to artistically tell a story, we can convey more meaning through sound and light.

Multimedia Interaction

Because we can use digital media interactively on a CD or on the World Wide Web, viewers can navigate through information sites, following links that satisfy their interests and needs. In effect, viewers create their own story structure but need cognitive maps to be able to navigate. Cognitive maps should have a home page, hub, portal, or gateway to serve as a starting point. There also should be clear links or paths people can select and move through. As with a good physical environment, there should be nodes that provide a sense of place. A good information environment should have clusters—much like districts and neighborhoods in a community—along with landmarks and focal points to add interest and aid in navigation. There may also be information "firewalls" or barriers between different domains through which those who are authorized can enter using passwords or other identification that functions like keys. Unauthorized hackers have emerged on the Internet like intruders or burglars in the physical world. And unfortunately some people produce "viruses" and other evil agents that are like bombs and terrorists in the physical world. The information environment we are creating on the web reflects both the good and bad of humanity and is becoming an expression of a new culture.

Developing cognitive maps is important to creating both CDs and websites. The CD accompanying this book provides a home page that identifies key needs and goals. (There is a goal related to each chapter of this book.) A cognitive map and description of how to navigate the CD is on page xi in the preface of this book.

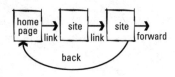

Cognitive map

In much the same way that architects of physical environments can help people find their way through the careful design of real space, authors—the architects of interactive multimedia environments—can organize information in ways that help people develop cognitive maps to orient themselves in cyberspace. Many issues such as imaging, place-making, way-finding, as well as community and privacy apply to both physical space and cyberspace. The classical education of many design professionals addresses these issues and can help us become architects of information environments. Considerable design research needs to be done on how people can more successfully navigate information environments to carry out activities ranging from e-commerce to education. We are just beginning to create a whole new world that will hopefully become inviting and worthwhile. It's success depends, to a great extent, on those who design these interactive multimedia environments. There is potential to create information environments that will extend the horizons of communication and knowledge.

Media, Models, Methods

There are three basic considerations when approaching multimedia production:

• Selecting media for transferring information and setting up the information environment

• Developing effective models so we can represent what we are working with digitally.

• Applying appropriate methods for interacting with these models using available digital tools.

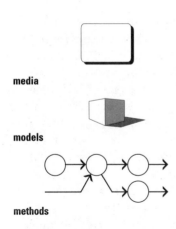

media

models

methods

Media, models, methods

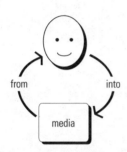

from   into

media

Multidirectional

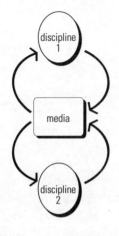

discipline
1

media

discipline
2

Multidisciplinary

When selecting media, we need to address the following questions: How does multimedia relate to our organizational goals? What is the message and how are we going to treat it? Who is the audience? How can this audience access shared information? Where do we store it? When making presentations, what is the story structure, script, and storyboard? When creating an interactive information environment, what is the cognitive map of the shared information environment? How is the information environment organized and information delivered? What are the inputs and outputs and what is the flow of information using technology we have access to? What is the best way to get the message to the intended audience?

When developing models we need to address the following questions: How can we most effectively model the realities we are dealing with? How can we share access to these models and interact with them? How can all the members of our workgroup relate to these virtual realities and add value by developing them? How can we most effectively mockup the multimedia production? How do we transfer multimedia information from our models to the reports, presentations, or interactive information environments we produce?

When applying appropriate methods, we need to address these questions: In preproduction, what treatment resolves the issues related to how we use media? In production, how do we develop the models? What tasks need to be done? Who is responsible for doing what? What is the schedule? In postproduction, how do we package the multimedia report, presentation, or information environment? What is the most effective workflow to carry out the preproduction, production, and post-production using multimedia within the available time frame and budget? How do we coordinate activities and events and optimize the team's efforts to achieve excellent results?

These approaches are multidirectional. They address getting information into electronic media as well as getting information from it. They can make it easier to organize and manage the creative collaboration of project teams using digital multimedia. There are many ways to work with multimedia using the digital tools that are available. Although these tools are rapidly evolving, basic approaches and work patterns related to thinking through these fundamental questions help guide approaches we can refine for each specific project.

These approaches are multidisciplinary. They are useful for focusing the creative thinking involved in many disciplines—ranging from the arts, to all the design professions, to business, and even to the sciences. They can help people use multimedia digital tools appropriately and combine expertise in effective team effort.

217

Transforming Media

transform *To change the form or condition of something. For example, digital tools can transform information from paper to electronic media, or from one file format to another.*

input-output matrix *A chart relating "what comes in" to "what goes out." Can help people consider all possible connections.*

Because there are so many different formats and devices for working with digital media, we often need to develop clear strategies for transforming it onto files we can use in our work environment. A simple input-output matrix helps us understand how to transform information. We can do this by identifying the format for receiving information, identifying how we transform it, and identifying our output. A matrix, like the one shown to the left, can help examine each channel of information we are using. For example, we may have a photograph we want to work with. With a scanner we can transform it into a digital graphic format. The digital format to use really depends upon the graphics software we want to work with in the next step. We can examine key transformations, anticipating needs in this way. An input-output matrix also provides ways to examine how to integrate media so that we can determine the most effective way of using the digital tools we have access to.

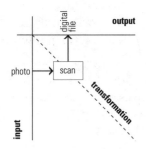

Input-output matrix

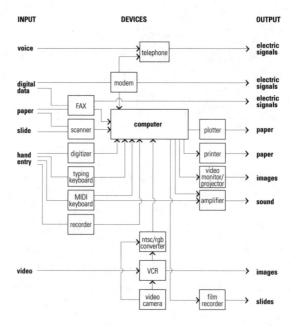

Input-output diagram

Formats

Electronic media provide both analog and digital information. Analog replicates sound or images using electrical impulses that modulate current. Digital has simple on-off signals, usually transmitted electronically. Audio and videotapes currently use electrical analogs of sounds and images. Television also currently transmits analog images, but is transitioning to HDTV and digital signals. CD and DVD players, as well as computers, use digital formats. Today the trend is toward digital formats because of the potential for higher quality and the possibilities for working more interactively with multimedia.

analog *That which corresponds to something else. For example, in electronic media, a format for replicating sound or images using electrical impulses that modulate current.*

digital *Relating to that which uses a binary system to replicate information using simple on-off signals, transmitted electronically or through fiber optics. Can represent numbers, text, graphics, images, video, and sound.*

raster image *Computer graphic composed of a bit map indicating which pixels to activate on a computer screen. Sometimes called a bit map image. Enables people to manipulate contrast, color, and texture.*

vector image *Computer graphic developed using algorithms that generate the geometry you see on the computer screen. Enables people to manipulate geometry — rotating, mirroring, and changing their scale.*

Digital graphic images can be either raster or vector. Raster images are bit maps indicating which pixels to activate on a computer screen. Vector images are developed using algorithms that show where vectors (lines) begin and end on a screen, along with their attributes. We can convert vector graphics to raster images by simply capturing the image on the screen and saving it in a raster graphic format such as TIFF. Converting raster images into vector requires digitizing by drawing over them with vector graphic tools and saving them in a vector graphic format, such as DWG. We also could use conversion software; however, conversion from raster may produce rather large and fragmented vector graphic files. Some graphic formats, such as EPS, handle both raster and vector information. Some formats, such as JPEG, also compress the digital data and are popular for interactive multimedia, especially on the web where small file size is important for speed.

We can work interactively with raster images, manipulating contrast and color. Raster-based programs simply address pixels on the screen, which change color to create an image. Imaging programs, such as Adobe Photoshop, are raster-based. The advantage of raster graphics is that it can more easily handle colors, textures, and tones—like those in photo images—or paint with an "electronic spray can." Both file size and resolution of raster images depend upon the number of pixels. Pixels are the color cells on a monitor or dots of color on a printed page. The higher the resolution the higher the number of pixels, as well as the larger the file size. Output devices may limit the resolution we see. This depends upon the number of dots per inch that the display will show. Most computer monitors have a resolution of 72 to 96 pixels per inch. Laser printers typically have a resolution of 300 dpi. Printing presses have higher resolutions. For example, an 8x10 inch photo scanned at 150 dpi would result in an 800 x 1000 pixel image that would print out as an 8x10 image at 150 dpi, or a 4x5 image at 300 dpi for use on monitors. Never enlarge images on the screen because that will reduce the resolution, unless the image has a larger number of pixels than is needed for the display.

Legibility is a related issue that we need to pay attention to when projecting words on digital screens for public viewing. We should use the 7x7 rule— 7 words per line and 7 lines per image. On monitors that we read up close, we can use about 12 words per line and 20 lines per screen. Monitors also have different aspect ratios that deal with the number of horizontal and vertical pixels. The current standard monitor has a frame of 640x480 pixels. D-1 NTSC has a pixel aspect ratio of 720x540. For DV, a recommended target size is 720x480, which is a wider format. Both the resolution of input devices, such as digital cameras, and output devices, such as video projectors and monitors, are increasing. Compression and higher-speed connections are making it possible to work with multimedia using higher resolutions.

Vector images use the number of pixels available in the display. Illustration programs, such as Adobe Illustrator, are vector-based programs, identifying the coordinates of points on lines that define the geometries shown on the computer screen. The advantage of vector graphics is that we can scale images without losing resolution. We can also plot drawings without the "jaggies" (or aliasing) associated with raster graphics. Vector programs have evolved many file formats; however, filters now permit transfers between most vector graphic programs' standards, with DXF—the data exchange file being an almost universal standard. Using many CADD and GIS programs, we can attach attributes to vectors and geometry to create "smart drawings" integrated with databases that can include multimedia.

Using animation we can create a series of images to create movement. Currently video works at 30 frames per second, but may change to 24 frames per second to reduce the file sizes of digital video and make video more consistent with film, which has traditionally worked at 24 frames per second. Some computer animation programs use the traditional "onionskin" metaphor for drawing a series of images to create smooth-flowing characters. Other animation programs use 3-D models that are manipulated to create movement or spatial sequences. We can coordinate sound tracks with the animated sequences and transfer them to videotape for presentation.

Digital video, using Quicktime or Media Player, enables us to view video presentations on a computer. Using digital video editing, we can speed up or slow down movement and create transformations and transitions in a nonlinear manner. This metamorphosis can result in the seemingly magical special effects we see in movies and on television. As HDTV emerges, the resolution of video images is improving. Although electronic images are not photographic quality, they are achieving resolution acceptable for publication and presentation. We can work with digital TV by capturing and manipulating single images as still video. We can also record and manipulate full motion digitally, although it is necessary to use data compression to keep these files at a workable size. Interactive video enables us to transmit video images in real time, creating the opportunity to set up telecommunication conferences that use live video.

Beyond integrating voice messages into e-mail and using voice recognition software, we may compose and record music using MIDI (musical instrument digital interface) keyboards. Digital synthesizers can even generate sounds, ranging from those of traditional instruments

digital video *Still or full motion images composed of binary information processed by computers. Enables people to speed up or slow down movement and create transformations and edit them nonlinearly.*

digital sound *Sound composed of binary information played by special consumer devices and computers. Enables people to change pitch and rhythm, mask noise, and mix sound tracks.*

Mix Media

to electronic music and sound effects. Audio production programs enable sound designers to manipulate digital sound—changing pitch and rhythm—masking noise, mixing sound, and adding other channels to a soundtrack or musical composition. Using multimedia software, we can integrate sound tracks into videos or other presentations.

Expectations rise as better tools become available. We often find ourselves having to meet others' expectations as well as our own. The web is like an open door. The composition we see on web pages cries for interesting formats. Video and animation seem like the old silent movies without sound tracks. By using digital tools we can create 3D models of whole environments. With more realism within reach, we can work in virtual reality. There are many exciting ways to use multimedia digital tools for design communication. There are also new opportunities to collaborate with other people online who have expertise to use different tool sets. We can even reach service bureaus electronically and engage them as if they were part of our organization.

By creating object-oriented workgroups we can use digital tools to mix media. We can produce elaborate printed reports and documents, mixing text and graphics. We can produce presentations simply using presentation software like MS Power Point, or more elaborately, using multimedia software like Director. We also can produce information environments and post them on web sites or distribute them on CDs or DVDs, engaging the audience in truly interactive ways. There are interesting differences between passive viewing and active exploration. For example, television tends to provide passive viewing, whereas video games can involve active exploration. Considering this dichotomy at more sophisticated levels, we can understand the differences that exist between a presenter/audience relationship and collegial collaboration or a workgroup relationship. This has tremendous implications for productivity, communication, and education. In general, the more people are involved, the more productive they are, the better they communicate, and the more they learn.

While multimedia production has been done in specialized studios, hardware and software are emerging which can empower design professionals to interactively use their perception right at a desktop. Putting multimedia phenomena into digital formats enables us to store, manipulate, and transfer these phenomena in ways that were never possible before. As electronic multimedia tools become easier to work with, they empower design professionals to produce more. Creative thinking and sensitivity can result in truly artistic expressions. People learning to use multimedia can move beyond simply learning techniques. As with any artform, it can take a while to discover the true qualities of these new media.

Summary
Strategies for Using Multimedia

Media - for transferring information

1. **Relate multimedia to your goals.** Develop a treatment describing who your audience is, what your message to them should be, and how you can reach them most effectively.

2. **Establish your information environment** and make sure there is the capacity to transfer and store multimedia files.

3. **Diagram information flow.** Examine input and output, looking for ways to capture multimedia directly using cameras, recorders and other devices. Seek the most effective ways to transfer multimedia into your report, presentation or interactive information environment.

Models - for addressing the realities involved

4. **Identify the realities you are working with** and determine the most effective multimedia models to use to explore and communicate those realities.

5. **Develop models of reality** producing the content for your report, presentation, or interactive information environment.

6. **Mock up your multimedia production.** Develop a working outline for your report, a script and storyboards for your presentation, or a cognitive map for your interactive information environment.

Methods - for refining the thought processes

7. **Identify what needs to be done** for preproduction, production and post-production.

8. **Diagram workflow.** Build a team, allocate responsibilities and determine the schedule.

9. **Manage progress.** Clarify and refine your procedures, improving what you do as you proceed.

Activities

Produce Multimedia

Design professionals work with multisensory phenomena. The challenge is to use multimedia to communicate more facets of a design in ways that audiences comprehend more readily. Take a project that you are currently working on and reconsider the media, models and methods you are currently using. Use the approach described here to develop better ways to use multimedia.

Initially you may look for ways to capture multisensory phenomena using tools such as digital cameras and recorders to help take field notes so you can work with this phenomena as you design. Think of it as doing digital sketching or painting using a digital camera and imaging software. Learn to incorporate mixed media into your current applications and use them for collaborating. For example, attach images or sound clips to e-mail. Integrate images of products into drawings; produce a video clip of your field notes. Make sure your team members have the capabilities to receive the multimedia information so they can learn to use it productively and appreciate its value.

Ultimately, learn to produce multimedia reports and presentations, as well as interactive information environments. This may involve engaging a team with the capabilities needed to carry out professional-quality multimedia production. Integrate images, drawings, digital models of all types, as well as animation, video, and audio. Put them on CDs, or DVDs, as well as on the on the web to reach your audience as the bandwidth of telecommunication increases. Use these multimedia productions for info-commercials, to document events, to explore design phenomena, to provide instructions for implementation, or even for motivational pieces or entertainment. Multimedia can open many new channels for design communication if we develop productive approaches for using it.

21

Ride
Waves of
Change

Design professionals can ride waves of change and not be drowned if we learn to design with digital tools and use new media creatively. The design professions are part of a new renaissance relating the arts and the sciences and influencing design. During the European Renaissance 500 years ago, civilization crossed a threshold of literacy. Many people gained access to the mechanical tools, such as pens and printing presses, and learned to work with traditional media, such as paper. As a result, creativity flourished and became manifested in what we now call the Renaissance. Today, civilization is crossing yet another threshold to a digital age. We now have access to digital tools that enable us to work more creatively with electronic media.

Why Change?

Key issues compelling change involve economic survival, environmental sustainability, and human fulfillment:

Economic survival in an information society requires learning to work with information technology. We need to navigate information environments, move data effectively, and conduct e-commerce. To enhance our own economic potential in a digital age, design professionals can begin by learning to use digital tools to set up information environments and access an amazing array of information services that are emerging, as well as online information, adding value through the design content we provide. We need to be able to build models of the realities we are working with to explore ideas, test alternatives, and communicate designs we can implement effectively. Automating some repetitive tasks can save time, reduce costs, and implement designs more quickly, linking computer-aided design with computer-aided manufacturing. But the real opportunities are to use information technology to improve quality, increasing the value we provide through collaboration and communication-aided design. There are also new opportunities to link design services with the market channels of new enterprises emerging on the World Wide Web.

Environmental sustainability requires recognizing that we are part of a larger environment that is our source of sustenance and inspiration.

Digital tools enable design and planning professionals to gather information about the environment on which to base better decisions regarding land use and resource management. Digital tools enable us to produce stronger regenerative designs fulfilling visions such as those articulated by John T. Lyle in the book *Regenerative Design for Sustainable Development*. We can model material and energy flows, as well as the effects of developmental actions, evaluating potential outcomes before committing resources. Although the digital technology enterprises involve using resources, such as electricity, this technology can help develop sustainable resources and renewable energy. Monitoring and managing complex operations with more precision can produce less waste, helping to sustain the ecosystems upon which we depend for life. We can use virtual reality to model the built environment, and provide clients and other stakeholders with a better sense of what they are investing in. Using information technology for more effective collaboration, each discipline can contribute its expertise, and help to sustain a healthy and stimulating environment.

self-actualize *To release your inner needs and potential. Psychologist Abraham Maslow places self-actualization at the top of a hierarchy of human needs.*

Human fulfillment provides possibilities for people to reach their creative potential—what Maslow calls self-actualization. There can be more to life than survival. Curiosity and creativity are strong motivating forces, enabling us to find out more about ourselves and about the world. New media provide many avenues for exploration and expression and set up opportunities for people to interact with each other, as well as learn more about the world around them. Digital tools can enhance the way design professionals perceive, think, and act in ways that can help cultures reach new levels of accomplishment. It takes creative people awhile to discover the qualities of a new media and the confidence and conviction to work with these qualities. It may take a culture longer to accept these qualities and embrace them aesthetically. But, exciting new images and art can emerge that have value, enabling people to explore the creative potential. Electronic media provide access to entertainment and education. Multimedia enables us to use almost all of our senses, involving many modes of thought and communicate through many channels of expression.

Some change is externally driven by technologies that offer better ways of doing things. Deeper changes are internally driven by shifts in mindset or attitudes that enable people to experience personal and spiritual growth. Stèphano Sabetti's book, *Waves of Change: Dynamics and Practice of Personal Change*, addresses the contents, theory, disturbance, and practice of change. Through his writing and the Institute for Life Energy, Sabetti, offers guidance in addressing deeper levels of change through an approach which uses energy vibrations, enabling people to experience more direct contact with their whole beings.

225

Risks and Rewards

There are both risks and rewards when embracing new technologies. We only need to consider the impact of the automobile, or television, to realize this. Automobiles provide welcome mobility. Yet indiscriminate use of automobiles in urbanized areas results in gridlock and environmental pollution. Television provides access to information in a compelling multimedia format, yet indiscriminate programming of TV results in appalling drivel that may pollute the minds of children.

Although we may gather information in virtual reality, it is essential to stay in touch with reality. Actually, as a landscape architect, I find that exploring electronic sources of information can enhance exploration of the real world, particularly when I can get involved interactively. We don't need to choose between virtual reality and nature. You can explore both the virtual reality of cyberspace and the reality of nature.

Abuses of the World Wide Web serve as a reminder that the sinister side of human nature can also enter virtual reality in the form of viruses and pornography. But virtual reality can also be used to create a world that might, in some ways, be better than the real world. And virtual reality can be used to visualize ways of overcoming real-world problems.

There are also dangers of global enterprises creating information monocultures—where major portions of the information ecosystem are subject to the same viruses, or Y2K programming glitches. And what would happen to design firms if the power went out in an earthquake, hurricane, or terrorist attack? Could we mobilize to help communities, or would we be casualties—not having adequate backup? Obviously, we need to think more about distributed computing and distributed power production, linking sustainable enterprises into rich and diverse information ecosystems.

Waves of Change

There are different ways to catch the waves of change. We can leverage investments in information technology to reduce operating costs of our business practices, or add value to the way we practice.

The challenge is to adopt information technology in ways that will get beyond the trade-off point when comparing the cost of e-business to the cost of standard business. At the same time, if we can also get beyond the transition point—where what we are doing with e-business is more valuable that what we are doing with standard business practices—we will catch the waves of change.

Early adopters may miss waves of change and lose money if what they endeavor to do with information technology ends up costing more and not increasing the value of standard business practices. Some of the early adopters of CAD and early websites provide examples of this.

Just-in-time adopters may find a period where, by reducing costs and adding value, they have a competitive advantage over standard practices, but unless they keep progressing, they may lose the wave to competition, as the innovations become standard practice. That is now occurring in the design professions with CAD.

Consequently, design firms need to continue to innovate, seeking the thrust of the wave of change to provide profits that can propel continuous improvement. But they also need to be careful not to be too far ahead, where there may not be a return on investment, or too far behind, where it may be a struggle to catch up. Design schools also need to stay in front of the crests of the waves of change. They should become not just training centers for current technology, but also breeding grounds for innovation.

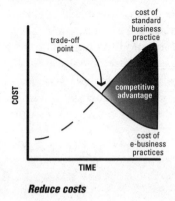

Reduce costs

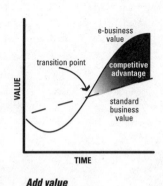

Add value

Shaping the Information Society

Until recently, and particularly during the industrial age, there has been a tendency to use materials and energy more readily than information. For example, it has always seemed easier just to get more fuel rather than use information to develop more efficient ways to use fuel. Historically, people have typically used empirical procedures—trying things out and testing them in reality—rather than optimizing designs with models before choosing which to implement. While empirical testing is useful, it can result in excessive expenditures of limited resources.

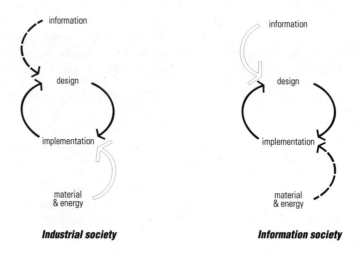

Industrial society   **Information society**

Design professionals can optimize designs using digital models. More information can create designs that will use less material and energy—particularly over a design's life cycle. It takes information to understand the result of extracting resources and releasing emissions. It takes information to understand how to integrate human life support systems into the natural process of the planet in a sustainable way. Digital models enable us to visualize better alternatives. This can be a catalyst for constructive change. In this way we can nurture the will needed to develop ways for securing and sustaining a more certain future.

Regenerative planning and design that reduces the consumption of material and energy could be at the core of information societies. Information technology could help us shift to resources that are more sustainable. An information society could lead to international cooperation where countries could share information to reduce demands on the planet's resources. Such a society could communicate ways to protect and enhance the environment, improving the quality of life. An information society could also lead to more sustainable agriculture in each region and better distribution of food to support the world's population. It could also disseminate information on birth control to help stabilize population growth and information to improve public health through prevention of disease. Freedom to access information is the basis for democratic forms of government.

As present-day society evolves into an information society, there can be a new integration of art and science. Art uses information in more expressive or evocative ways. Science uses information in more rational or logical ways. Each draws upon different modes of thought. Digital multimedia enables people to work with all these modes of thought.

Hopefully we will use digital tools in wise and ethical ways. Careful design of our information environment can help sustain a healthy economy. Design professionals can help shape the information society in the new millennium. We can begin by changing our work environment, drawing more upon online information resources. We can change our work patterns to use digital tools and shared information environments more effectively. We can also find constructive ways to change the organizational structures, using technology to transfer and transform information more effectively. Productivity can improve by focusing on information objects and information transfers—not on an organizational structure based on a hierarchy of positions. An organization based on flexible workgroups, rather than a rigid hierarchy, lends itself to finding the most effective ways to organize and move information. Workers no longer have to settle for "the way it is done" but rather can find "the way that makes sense."

A New Renaissance

New technology is setting the stage for a new renaissance, potentially as significant as what occurred 500 years ago. While we can see the driving forces of change at work, what happens will involve content providers who shape our culture. These include design professionals—artists, architects, planners, engineers, and many others—creatively using the tools being developed by innovative people in information technology.

The French historian Jules Michelet coined the term "Renaissance" in 1855 while writing about what occurred in European culture during the fourteenth, fifteenth, and sixteenth centuries. He said the Renaissance was the "discovery of the world and of man." Today, we are witnessing a discovery of the information world and, hopefully, humanity.

Movable type from print by Jost Amman

If we examine the Renaissance, we see there were fundamental shifts in the way people addressed basic needs such as communication, collaboration, and commerce. Again, today, we are witnessing remarkable changes.

Communication changed during the Renaissance when movable type made possible a proliferation of mechanical printing presses. This enabled people to print documents, distributing knowledge on paper. Now, digital media are bringing about the convergence of telephone, FAX, film, radio/audio, TV/video, and computers in ways that enable people to interact with digital multimedia online.

Collaboration changed during the Renaissance when more people learned to get information, thoughts, and feelings onto paper. There was a proliferation of models of reality—such as drawings, maps, diagrams, and mathematical formulas. Disciplines built tremendous knowledge bases. Today people are learning to interact using digital models and to collaborate in shared information environments. The knowledge base continues to grow at exponential rates and is becoming more accessible online.

Five hundred years ago, the availability of new pigments, oils, and materials for frescos stimulated the art world. Today, we have interactive, animated multimedia available in millions of colors and rich sound.

During the Renaissance, there was the discovery of linear perspective. Today, we can work with digital 3D models in virtual reality. At that time, drawing and painting demonstrated better ways to render light, and shadow. Today, we have programs with ray tracing that can render luminosity. And animation programs enable us to work with the fourth dimension—time.

Linear perspective from print by Albrecht Durer

Commerce also changed dramatically during the Renaissance. New regional economies emerged, driven by city-states using advancements in transportation to open new trade routes. The concentration of wealth shifted from landowners to investors and merchants. The Church became less of a dominant force as individuals gained more control over their destiny. Today, we are seeing a global economy emerge, driven by nations having access to new communication and transportation technology. E-commerce is handling a growing number of transactions online. And there is a shift in wealth from industrial-based entities to new information-based entities. Large organizations may become less dominant forces as new technology empowers individuals.

Advances in transportation during the Renaissance began making it easier to move people and goods, thereby changing people's perception of space and time. Today, with the emergence of information superhighways, we are exploring information environments and moving information at the speed of light. Again we must rethink how we relate to space and time. Telecommunications enable us to move information and ideas sustaining human interaction in ways not conceivable before.

Drawing by Leonardo da Vinci

During the Renaissance there was an integration of art and science as creative people like Leonardo da Vinci used detailed drawings and scale models, as well as mathematical formulas, to explore and express ideas. Today, art and science merge again through imaging and scientific visualization. We can see fractal images and work interactively with very complex, even chaotic, visual simulations. We can draw upon more modes of thought as we use these new models to represent the realities we are dealing with. Creative content providers are changing their mindsets and adopting web workstyles.

Education

To realize the full potential of this new renaissance will require advancements through education. Media influence how we use the creative process. Working with paper is somewhat different than working with film, and working with film is different than working with electronic media. Each has intrinsic qualities that we can use creatively. A major challenge today is to learn to use electronic media creatively.

Learning curves

Changes in thinking skills are slower than technological changes. It takes years for some people and societies to accept new tools. In addition, people may need several more years to master complex applications. Those who are excited about the possibilities will start to change their approaches. Change begins with an awareness of new tools, but it takes a desire, and commitment, to learn to use them creatively. This requires time and patience.

Education is changing people's mindsets within the context of colleges and universities. It also is occurring through training and continuing education programs, as well as through the self-development of dedicated individuals.

An important mission for education is to help people develop their mental capacity. This requires focusing on creative, as well as critical, thinking skills. Developing thinking skills enables people to continue learning to use tools and reach their full creative potential. User training for specific hardware and software often becomes dated within a few years; however, it can be a foundation for further training.

There are three general approaches for learning to work with digital tools. One approach is to learn by doing it. Another approach is to learn about it. And yet another approach is to visualize it. We can see some aspect of these approaches used in almost any school or training program.

Training courses are best for learning techniques. Often what is necessary to get started uses complex hardware and software. However, there is a danger of losing sight of what we can really do with these tools. Some students feel that what they are doing is gaining marketable skills; however, there is more to life than getting an entry-level position as a computer operator; training can quickly grow stale. It may also become obsolete when the software or hardware is updated or superseded. Many educational institutions have difficulty maintaining up-to-date training facilities. The better training facilities are often in training centers and vocational schools.

Survey courses can provide background knowledge about computers and related topics. Traditional learning focuses on terminology and techniques before expecting someone to use tools creatively. This broader understanding, for example, provides an appreciation of what has gone into the development of these amazing tools and a sense of where the computer industry is moving. The problem is, students may know about everything, but be unable to do anything. Even worse, they may not even visualize their potential or imagine what they might be able to do with these tools. They haven't learned that answer because they haven't developed their capacity to think creatively.

Courses teaching thinking skills can help people learn to use digital tools creatively. These courses help people visualize how to use electronic multimedia in their disciplines or areas of interest. They enable people to play with and picture possibilities as well as learn terminology and techniques related to our information environment. We can also use visualization techniques for self-study, develop our own inner game, even without having access to the latest digital tools. We are all prisoners of different sets of circumstances, but that should not limit the possibilities we can envision. Design professionals who develop their thinking skills can grow and flourish as they gain access to better digital tools. And as we become more comfortable with new media by working intimately with it, we can adopt web workstyles and lifestyles.

"So far as a man thinks, he is free." --
Ralph Waldo Emerson
The Conduct of Life: Fate (1860)

The whole notion of using shared information environments for object-oriented learning and workgroups has great potential. Sometimes the most interesting approaches for using computer applications come from those who have not learned so-called standard procedures. New approaches can come especially from those who don't know what, supposedly, can't be done. Those who focus on user training may perfect certain procedures. The challenge is to go beyond these procedures and develop the thinking skills to grow, riding waves of change with the new tools that are emerging.

Riding the waves of change, we will be able to interact and collaborate more productively online, drawing upon the creative capacity of multi-disciplinary teams to address important environmental problems. We be will be able to manage projects efficiently, acquiring and working more effectively with valuable information. We will be able to implement better ideas quickly. Through a full range of digital models—especially 3D models of virtual reality—we will be able to reach broader audiences with the splendor of multimedia.

Summary
Strategies
for Riding
Waves
of Change

1. **Identify with driving forces compelling change** that are meaningful to you. Especially consider economic survival, environmental sustainability, and human fulfillment when searching for the motivation to ride waves of change.

2. **Carefully consider both risks and rewards** to avoid pitfalls and realize the potential.

3. **Look for opportunities to reduce operating costs and increase value** by adopting digital information technology. Seek innovative strategies for continuous improvement.

4. **Consider timing.** Avoid adopting too early, or too late, by carefully considering the return on your investments in information technology. If it is too expensive, you may be too early. If it is too cheap, you may be too late to get a competitive advantage.

5. **Look for ways to use more information and less material and energy** in what you design and do. Use digital tools in wise and ethical ways that can help improve our information society.

6. **Realize that you are contributing to larger changes** which, when considered from a historical perspective, may become as significant as the Renaissance 500 years ago.

7. **Seek and support education** to help change mindsets enabling you and others to become more receptive to change and capable of seizing opportunities that are emerging. Develop both technical and thinking skills, as well as your capacity to envision the potential. Guide change constructively.

8. **Ride waves of change** by acting upon opportunities that arise so you can enjoy the benefits that can be derived from digital tools, a web workstyle and even a web lifestyle. Design with digital tools, using new media creatively.

Activities

Change Work Habits, Environments, and Organizational Structures

By effectively using information technology, design professionals may ride the wave of change and not drown. Anticipate change and see where waves of change are going. Timing is also important—not adopting new tools too early, or too late. Seek your creative potential and don't beome frustrated or limited in unnecessary ways.

The challenge is to set up new work patterns—to find more freedom over how we use space and time. Look especially for ways of doing more with fewer resources. Consider how individual and group work patterns could change once you establish shared information environments. Try out these changes using the environment you have access to. Begin with changing your personal work patterns. Then find others who are also making these changes so you can work together.

Changes in the work environment and in work patterns are also changing organizational structures. For example, there is potential to do work using telecommunications, reducing the number of seemingly endless meetings that fit into nobody's schedule and often have no tangible result. Working electronically can avoid trips that add to meeting time. It can save production time involved in delivering the results of collaborative efforts. A new sense of community can evolve related to cyberspace. One only needs to search the web to see how they link special-interest groups consisting of people from all over the world.

The challenge here is to consider our own organizational structure. Examine how it might evolve to respond to new work patterns that could result from the information environment we are shaping. Begin with pilot projects and consciously try new approaches. Evaluate the results; improve the approaches; then implement changes.

These final activities at the end of this chapter are really more challenges than exercises. They challenge us to use creative thinking skills to change our work environment, work patterns, and even organizational structures. The hope is that these changes could reverse some of the trends related to demoralization of humanity and degradation of the environment. The changes we make might help lead to more satisfying, creative, and environmentally conscious lives. Although changes are challenging, they could contribute to happiness and productivity.

235

Appendices

PART I   Methods

Design teams—including managers, design professionals, and staff—
need ways of working effectively. This part of the book offers methods
to use new media and digital tools creatively and productively for
design communication in professional practice. It can be a helpful text
in courses on design methods and new media for design communication, as well as computer courses that go beyond application training.

PART II   Case Studies

Good examples demonstrate how progressive design professionals are
using new media and applying digital tools in their practices. The case
studies examine a range of design practices—writing and graphic
design, creating multimedia and special effects, landscape architecture,
architecture, and planning. The case studies explore what works in
office environments as well as in virtual offices and electronic studios
using online collaboration.

PART III Strategies

Once we invest and commit to using new media and digital tools, it is
important to find ways to get the most out of them. This part of the
book is devoted to realizing the potential inherent in information technology. It will provide valuable help as your design team addresses
change.

APPENDICES

Bibliography

Adams, James L.: *Conceptual Blockbusting: A Guide to Better Ideas*, 2nd Ed., W. W. Norton, New York, 1979.

Adams, James L.: *The Care & Feeding of Ideas: A Guide to Encouraging Creativity*, Addison-Wesley, Reading, Mass.,1986.

Alexander, Christopher et al.: *Pattern Language*, Oxford University Press, New York, 1977.

Arnheim, Rudolph: *Visual Thinking*, University of California Press, Berkeley, 1969.

Brinkmann, Ron: *The Art and Science of Digital Composting*, Academic Press, San Diego, 1999.

Brooks, Rodney A.: *Model-Based Computer Vision*, UMI Research Press, Ann Arbor, Mich., 1984.

Bruner, Jerome: "The Conditions of Creativity," in *Consciousness: Brain, States of Awareness and Mysticism*, Scientific American, Harper & Row, New York, 1979.

Carroll, Lewis: *Alice's Adventures in Wonderland & Through the Looking-Glass*, (originally published in 1872) republished by Macmillan, New York,1966.

Csikszentmihalyi, Mihaly: *Flow: The Psychology of Optimal Experience*, HarperCollins, New York,1990.

de Bono, Edward: *Lateral Thinking: Creativity Step by Step*, Harper & Row, New York, 1970.

DeKoven, Bernard: *Connected Executives*, Institute for Better Meetings, Palo Alto, Calif., 1988.

Dertouzos, Michael: *What Will be: How the New World of Information Will Change Our Lives*, HarperEdge, New York,1997.

Dewey, John: *Art as Experience*, G. P. Putnam's Sons, New York,1934.

Doblin, Jay (ed.): *Design in the Information Environment*, Knopf, New York, 1985.

Edwards, Betty: *Drawing on the Artist Within*, Simon and Schuster, New York, 1986.

Edwards, Betty: *Drawing on the Right Side of the Brain*, J. P. Tarcher, Los Angeles, 1979.

Emerson, Ralph Waldo: *The Conduct of Life: Fate*, Houghton Mifflin, New York, 1904.

Fezler, William: *Creative Imagery: How to Visualize in All Five Senses*, Simon and Schuster, New York, 1989.

Franck, Frederick: *The Zen of Seeing*, Random House, New York, 1973.

Galluzzi, Paolo (ed.): *Leonardo da Vinci: Engineer and Architect*, Montreal Museum of Fine Arts, Montreal,1987.

Gates, Bill: *Business @ The Speed of Thought: Using a Digital Nervous System*, Warner Books, New York,1999.

Gawain, Shakti: *Creative Visualization*, New World Library, Novato, California, 1995.

Getzel, Jacob W.: "The Psychology of Creativity," *Carnegie Symposium on Creativity*, Library of Congress Council of Scholars, Pittsburgh,1980.

Gibson, William: *Mona Lisa Overdrive*, Bantam Books, New York, 1988.

Gibson, William: *Neuromancer*, Berkley Publishing Group, New York,1984.

Gleick, James: *Chaos: Making a New Science*, Penguin Books, New York,1987.

Gordon, William J. J.: *Synectics: The Development of Creative Capacity*, Harper, New York,1961.

Halprin, Lawrence: *RSVP Cycles: Creative Processes in the Human Environment*, Braziller, New York,1969.

Halprin, Lawrence, and Jim Burns: *Taking Part: A Workshop Approach to Collective Creativity*, MIT Press, Cambridge, Mass., 1974.

Hanks, Kurt, and Larry Belliston: *Draw! A Visual Approach to Thinking, Learning and Communicating*, William Kaufmann, Los Altos, Calif., 1977.

Hanks, Kurt, and Larry Belliston: *Rapid Viz: A New Method for the Rapid Visualization of Ideas*, William Kaufmann, Los Altos, 1980.

Harder, Christian: *Serving Maps on the Internet: Geographic Information on the World Wide Web*, ESRI, Redlands, California, 1998.

Heller, Steven, and Daniel Drennan: *The Digital Designer: The Graphic Artist's Guide to the New Media*, Watson-Guptill, New York, 1997.

Henri, Robert: *The Art Spirit*, J. B. Lippincott, Philadelphia, 1923.

Herrmann, N.: *The Creative Brain*, Brain Books, Lake Lure, North Carolina, 1989.

Hirschtick, Jon K.: "A Solid Future Is Emerging", *Desktop Engineering*, March 1999, page 80.

Hirsh, Sandra K. and Jean M. Kummerow, *Introduction to Type in Organizations*, 2nd ed., Consulting Psychologist Press, Palo Alto, California, 1990.

Hofstadter, Douglas R.: *Gödel, Escher, and Bach: An Eternal Golden Braid*, Random House, New York, 1980.

Huxley, Aldous: *The Doors of Perception*, Harper & Row, New York, 1954.

Jaynes, Julian: *The Origin of Consciousness in the Breakdown of the Bicameral Mind*, Houghton Mifflin, Boston, 1976.

Johnson, Robert A.: *Inner Work: Using Dreams and Active Imagination for Personal Growth*, Harper & Row, New York, 1986.

Jung, C. G.: *Man and His Symbols*, Doubleday, Garden City, N.Y., 1964.

Kneller, George: *The Art and Science of Creativity*, Holt, Rinehart and Winston, New York, 1965.

Koberg, Don, and Jim Bagnall: *The Universal Traveler*, William Kaufmann, Los Altos, Calif., 1976.

Krueger, Myron W.: *Artificial Reality*, Addison-Wesley, Reading, Mass., 1983.

Laiserin, Jerry: "A New Kind of CAD—Communication-Aided Design,"*Cadence*, June 1999, pp.18-26.

Laseau, Paul: *Graphic Thinking for Architects and Designers*, Van Nostrand Reinhold, New York, 1980.

Lowenfeld, Viktor: "Basic Aspects of Creative Thinking," in *Creativity and Psychological Health*, M. F. Andrews (ed.), Syracuse University Press, Syracuse, 1961.

Lyle, John: *Design for Human Ecosystems*, New Edition, Island Press, Washington, D.C., 1999.

Lyle, John: *Regenerative Design for Sustainable Development*, John Wiley & Sons, New York, 1993.

Lynch, Kevin: *Image of the City*, M.I.T. Press, Cambridge, Mass., 1960.

Maslow, Abraham H.: *Motivation and Personality*, Harper & Row, New York, 1970.

Maslow, Abraham H.: *Toward a Psychology of Being*, Van Nostrand, Princeton, N.J., 1968.

Maslow, Abraham: *The Farther Reaches of Human Nature*, Penguin, New York, 1976.

McGovern, Gerry: *The Caring Economy—Digital Age Business Principles*, Blackhall Publishing, Dublin, Ireland, 1999.

McKibben, Bill: *The Age of Missing Information*, Random House, New York, 1992.

McKim, Robert: *Experiences in Visual Thinking*, Brooks/Cole, Monterey, Calif., 1972.

McLuhan, Marshall, and Quentin Fiore: *The Medium Is the Massage*, Bantam Books, New York, l967.

Meyer, Chris, and Stan Davis: *Blur: The Speed of Change in the Connected Economy*, Addison Wesley, New York,1998.

Michell, William J., and Malcolm McCullough: *Digital Design Media*, Van Nostrand Reinhold, New York, 1991.

Minsky, Marvin: *The Society of Mind*, Simon and Schuster, New York,1988.

Mitchell, Andy: *Zeroing In: Geographic Information Systems at Work in the Community*, ESRI, Redlands, California, 1998.

Myers, Isabel Briggs, and Peter B. Myers: *Gifts Differing*, Consulting Psychologists Press, Palo Alto, Calif.,1980.

Neisser, Ulric: *Cognitive Psychology*, Prentice-Hall, New York, 1967.

Odum, Howard T., and Elisabeth C. Odum: *Energy Basis for Man and Nature*, McGraw-Hill, New York,1976.

Odum, Howard T.: *Environment, Power, and Society*, John Wiley & Sons, New York, 1970.

Ornstein, Robert, and Paul Ehrlich: *New World, New Mind: Moving toward Conscious Evolution*, Doubleday, New York,1989.

Rico, Gabriele L.: *Writing the Natural Way: Using Right-Brain Techniques to Release Your Expressive Powers*, J. P. Tarcher, Los Angeles,1983.

Rodriguez, Walter: *The Modeling of Design Ideas: Graphics and Visualization Techniques for Engineers*, McGraw-Hill, New York, 1992.

Sabetti, Stèphano: *Waves of Change: Dynamics and Practice of Personal Change*, Life Energy Media, Sherman Oaks, California, 1993.

Sabetti, Stèphano: *The Wholeness Principle, Exploring Life Energy Process*, Life Energy Media, Sherman Oaks, California, 1986.

Schon, Donald A.: *Educating the Reflective Practitioner: Toward a New Design for Teaching and Learning in the Professions*, Jossey-Bass, San Francisco, 1987.

Schon, Donald A.: *The Reflective Practitioner: How Professionals Think in Action*, Basic Books, New York, 1983.

Schrage, Michael: *Shared Minds: The New Technologies of Collaboration*, Random House, New York,1990.

Schulz, Robert C.: *CIMS Project Management Manual: Practice & Submittal Requirements for a Coordinated Information Management System*, Cal Poly Pomona (portion of masters thesis), 1997.

Schulz, Robert C.: *CMC in AED: Computer Mediated Communications in Architecture, Engineering, and Construction*, Cal Poly Pomona (portion of masters thesis), 1995.

Silva, Mira, & Shyam Mehta, *Yoga: the Inyengar Way*, Knopf, New York, 1995.

Sorkin, Michael: "The Electronic City," *I.D. Magazine*, June 1992, pp. 71-77.

Springer, Sally P., and Georg Deutsch: *Left Brain, Right Brain*, W. H. Freeman, New York, 1981.

Stauffer, Russell G.: *Teaching Reading as a Thinking Process*, Harper & Row, New York, 1968.

Stitt, Fred: *Model Working Drawing Sheets*, GUIDELINES, Orinda, California, 1999.

Stults, Bob: *Media Space*, Xerox Corp., Systems Concepts Laboratory Technical Report, Palo Alto Research Center, Palo Alto, Calif., 1986.

Taylor, Gordon Rattray: *The Natural History of the Mind*, Dutton, New York, 1979.

Tufte, Edward: *Envisioning Information*, Graphics Press, Cheshire, Conn., 1990.

Tufte, Edward: *The Visual Display of Quantitative Information*, Graphics Press, Cheshire, Conn., 1983.

Unterseher, Fred, Jeannene Hansen, and Bob Schlesinger: *Holography Handbook: Making Holograms the Easy Way*, Ross Books, Berkeley, Calif., 1987.

von Oech, Roger: *A Whack on the Side of the Head*, William Kaufmann, Los Altos, Calif., 1983.

von Wodtke, Mark: *Mind over Media: Creative Thinking Skills for Electronic Media*, McGraw-Hill, New York, 1993.

Welles, Edward O.: "The Perfect Internet Business," Inc. August 1999, pp. 71-78

Westbrook Adele, and Oscar Ratti: *Aikido and the Dynamic Sphere*, Charles E. Tuttle, Tokyo, 1970.

Whitehead, Alfred North: *Modes of Thought*, Macmillan, New York, 1938.

Wilber, Ken: *No Boundary: Eastern and Western Approaches to Personal Growth*, Shambhala Publications, Boston, 1979.

Woolsey, Kristina Hooper, Scott Kim, and Gayle Curtis: *VizAbility, Changing the Way You See the World*, PWS Publishing Company, Boston, 1996.

Glossary

A

abstraction An operation of the mind involving the act of separating parts or properties of complex objects. Enables you to simplify information and clarify relationships you perceive.

actual prototypes Three-dimensional physical models.

agents Bots (robots) or droids (androids) that help gather specific information.

aikido (Japanese) A martial art that focuses on three elements: *Ai* refers to harmony or coordination; *ki*, spirit or inner energy; and *do*, the method or the way.

aliasing Assuming the characteristics of a computer generated line. Associated with raster graphics where dots form lines by reducing "jaggies," or a steplike appearance.

analog That which corresponds to something else. For example, in electronic media, a format for replicating sound or images using electrical impulses that modulate current.

anastrophe A change involving the coming together of disparate elements to form a coherent and connected whole. The opposite of catastrophe.

animation program A computer application that enables users to portray movement.

anthropomorphic Related to humanlike forms or attributes.

applet A small application program that can be downloaded from the web.

archetype In psychology, according to Jung, a symbol of the inner being that manifests ideas inherited from human experience. These symbols recur in many different cultures over time.

artificial intelligence The ability to carry out programmed responses. Computers can carry out programmed responses.

artificial reality A model or representation of reality. People can develop this model using information in media space.

asynchronous At different times. Online communication engaging participants in interaction at different times, as in e-mail correspondence.

attribute A property, quality, or characteristic that describes an object.

avatar a virtual being that can cross over into cyberspace.

B

background plate (BG plate) The back plate for live action (often shot in front of a green-screen).

bandwidth A range. Multimedia encompass a wide bandwidth.

biofeedback Technique for responding to indicators of mental and body stress. For example, therapists use biofeedback to help patients learn how to relax and overcome physical and mental problems.

bonding Connecting or holding together. The formation of close interpersonal relationships involving the whole being. People can also bond with objects related to computers, and with places in reality or virtual reality.

bubble diagram A graphic representation that shows the relationships between functional areas. Can describe patterns evident in a map or a plan.

C

compact disk-read only memory (CD-ROM) A laser disk for storing digital information. Typical capacity is 650 megabytes.

cellular automata Computer programs enabling objects to carry out functions. Empower information objects to begin to take on a life of their own.

checklist A list of activities or items. Often called a "to do" list.

cognitive Knowing; recognizing what you perceive. Cognitive computer applications use primarily deductive modes of thought.

cognitive map A visual representation of what you know. A cognitive map can show spatial or conceptual relationships.

collaboration The act of working together. Electronic media offer new opportunities for workgroup collaboration.

computer-aided design and drafting program (CADD) A computer application that enables users to model and develop drawings of designs.

concurrence Happening together. Agreement; accord. Concurrence makes it possible for more people to work together with the same information.

contemplate To access different modes of thought through meditation. To reflect upon what you experience and understand.

CPM (Critical path method) A procedure for determining which set of activities will take the longest to accomplish, hence constituting the critical path of activities in a project.

creative process The stages your mind goes through when developing ideas. These stages include preparation (involving both first insight and saturation), incubation, illumination, and verification.

cyberspace Media space connected to the human brain, enabling people to experience this information environment interactively.

D

DAT Digital audio tape, often used for storing music.

data flow diagram A graphic representation that helps users visualize how to transfer digital files. Especially helpful when working with different file formats where it is necessary to figure out the best way to move files from one program or computer platform to another.

database program A computer application that enables users to sift information by selecting attributes.

delegate To entrust another person with a task.

design chain Enables a team of design professionals to add value ideas and information throughout the design process.

design phases The steps you go through when doing a design project. Each phase of design involves the design process.

design process The stages your mind goes through when developing design ideas. These stages — which are related to the stages of the creative process — include research (involving both problem identification and information gathering), analysis, synthesis, and evaluation.

desktop publishing program A computer application that enables users to lay out and integrate text with graphics.

desktop video program A computer application that enables users to create multimedia productions involving text, graphics, images, video, and sound.

dichotomy A division into two parts. A set of two (usually opposites).

digital Relating to that which uses a binary system to replicate information using simple on-off signals, transmitted electronically or through fiber optics. Can represent numbers, text, graphics, images, video, and sound.

digital community A group of people, with shared interests, who are linked though an online information environment.

digital nervous system Using information technology to extend our capabilities to perceive and respond.

digital office An organization that makes extensive use of digital tools.

digital sound Sound composed of binary information played by special consumer devices and computers. Enables people to change pitch and rhythm, mask noise, and mix sound tracks.

digital versatile disk (DVD) sometimes called *digital video disk* A high-capacity laser disk for storing digital information. A single-sided DVD can hold up to 4.7 GB on one layer and 8.5 GB on two layers. A two-sided, two-layer DVD has a total capacity of 17 GB.

digital video Still or full motion images composed of binary information processed by computers. Enables people to speed up or slow down movement and create transformations and edit them nonlinearly.

direct analogy A similarity or likeness that involves similar physical characteristics or processes that relate to different contexts.

directory A guide. The directory of a computer disk will provide a guide to its contents.

download To transfer digital information from a server (or mainframe computer) to a local computer.

drawing and painting programs Computer applications that enable users to draw and paint digitally.

dynamic data exchange The automatic transfer of information from one application to another as you work in any of the applications linked by the exchange.

E

e-business (electronic business) encompasses e-commerce, and includes applications to help business run more efficiently. It also includes more internal applications for linking employees together and helping employees work more productively. E-business also involves publishing and accessing information.

e-commerce (electronic commerce) E-commerce involves buying and selling (on the Internet) and all the processes that support buying and selling, such as advertising, marketing, customer support, and credit-card activities.

e-gov (electronic government) Providing governmental services and information online.

e-mail (electronic mail) Messages transmitted by computer to addresses on a server. These messages can be picked up anytime, from anywhere, using a computer to access the address.

ergonomics The study of how energy is spent. Pertains particularly to human energy expended for doing work.

ethic (From the Greek word *ethos*) The essential character or spirit of a person or people. The basis upon which people make unconscious judgments.

eXtensible markup language (XTML) Describes the content of a website using standard tags for links.

F

fantasy analogy A similarity or likeness that relates to an ideal.

file management utilities Computer programs that help users navigate around their information environment. Permit pruning of the branching structure of file directories and easily move branches and files.

file transfer protocol (FTP) The software code for uploading and downloading digital files.

flow diagram A graphic representation that shows the sequence of a process. Can describe movement of material or energy in natural processes.

G

Gannt chart A bar graph that shows the duration of tasks and relates them to a time line.

geographic information system (GIS) A computer application that enables users to develop spatial models with layers of information linked to databases of attributes.

gestalt, pl gestalten (German) The recognition of integrated patterns that make up an experience. The overall style or personality that one senses. This whole is more than a sum of the parts.

group dynamic Interaction among members of a team. Involves the members' different modes of thought.

groupware Software that enables group interaction Helps coordinate large teams using computers to do complex projects.

H

hardware The computer and its peripherals such as the central processing unit and monitor; includes input devices such as the keyboard, mouse, and microphone and output devices such as a printer, speakers, and modem..

hierarchy The governing or intrinsic order. Can be categorical as well as spatial and temporal.

high-definition television (HDTV) The new standard for digital television that has higher resolution and a wider screen than conventional analog color television.

hub An information site that provides access to the other sites.

human factor A characteristic related to people. Especially concerning how people interface with tools such as computers.

hypermedia A collection of keywords, graphics, images, video, and sound linked by associations. Used to present digital information in ways a user can explore interactively.

hypertext A collection of keywords linked to information. Used to present associated information so a user can quickly access what he or she is interested in, using a computer.

hypertext markup language (HTML) Defines the layout of text and graphics on web pages.

I

icon An image or representation used to express ideas graphically. Provides a landmark, helping computer users recognize where they have been, where they are, and where they are going in cyberspace.

image integration The combining or compositing of visual information. For example, draping air photos over computer terrain models.

imaging program A computer application that enables users to work with and enhance digital images.

information architects People who shape our information environment by mining data, organizing information for workgroups to share, and creating information sites where people can navigate and work with data they need.

information ecosystem Value chains of information feeding webs of enterprises online.

information flow diagram A graphic representation that helps users visualize how to transfer information into a digital format that they can work with interactively. Especially helpful when integrating information from many different sources and transforming it into a range of products.

information work Thinking work involving the transformation of information by human brains or computer programs.

inner space The mental realm that relates to human memory and spirit. Just as people can develop a mental map to navigate in media space, people can also develop an inner sense to navigate in their own inner space.

input-output diagram A diagram showing the paths through which material and energy (such as information) moves. Can help people visualize how computers, with the appropriate hardware and software, can transfer and transform information electronically.

input-output matrix A chart relating "what comes in" to "what goes out." Can help people consider all possible connections.

interact To view and do what you visualize.

interactive video Video that is both sent and received. Enables users to interact in real time involving both audio and video transmitted electronically.

Internet The public online environment that includes e-mail and other information services including the World Wide Web.

Internet protocol (IP) Standards for transmitting e-mail and other files on the Internet.

intranet An information environment established by an organization for internal use. Can be navigated using same tools used on the internet.

intrinsic order The natural or inherent order. Recognizing the intrinsic order of what you are working with enables you to model or represent it using a computer.

J

Java A cross-platform programming language for developing applets.

L

lateral thinking An approach to creative thinking typically involving consideration of alternatives.

life cycle A sequence of stages that include the formation, use, and disposal of an object. Can refer to physical objects as well as to information objects developed in computers.

local area network (LAN) Links computers and peripherals together so that they can share information within a location.

M

macro A brief computer program that combines steps. For example, a macro can make repetitive logging-on procedures easier when using computers.

mandala A video setting you walk into and interact with.

map To represent spatial order.

map of media space A visual representation of your information environment. Shows how to connect with and navigate in cyberspace. Can include what is accessible on disk, as well as what is accessible through networks and via telecommunications on the Internet and World Wide Web.

marketing channel Addresses how a product is conceived by its maker and received by its user.

media space The information environment connecting real and imaginary places and the people and objects within them. The environment in which people can use representations to work with virtual reality in cyberspace.

meditate A way of thinking that enables people to access altered states of consciousness. To project into unconscious modes of thought.

medium, pl media The intermediate material for expression. Using information technology, people work with ideas and information expressed in electronic media.

mental interface In computer applications, that which relates to software procedures and thinking skills.

menu A list of choices. Provides signs and pathways to help computer users find their way in their information environment.

metadata Data about data. Used by clearinghouses to help people find what information exists.

metaphor Transferring to one situation, the sense of another. For example, transferring to a computer application, the organization of a desktop. Helps you relate to previous experience upon which you can build familiarity.

mind That which thinks, perceives, feels, or wills; combining both the conscious and unconscious together as the psyche. The source of thought processes that facilitate the use of computers for artistic expression, design, planning, management, or other problem-solving and issue-resolving applications.

mindmap A diagram of associations showing how you link key words.

mindscape The inner world of your own mind, involving both conscious and subconscious levels.

mindset Attitude; point of view.

mockup A representation. Mock-ups show how the pieces of an object or the parts of a presentation come together. Enables visualization of a final product before producing it.

model To represent functional order.

model space The information environment where you work on your models.

mosaic A series of images. Used to preview a collection of digital graphic files so a user can select an image to call up on a computer.

multidirectional Involving more than one direction. Enabling computer users to both receive and send digital information.

multidisciplinary Involving more than one discipline. Enabling disciplines to work together.

multimedia Integrating more than one medium. Computer systems can enable the integration of electronic media combining text, graphics, animation, spatial modeling, imaging, video, and sound.

multitasking Using more than one software application at the same time. For example, some computer user interfaces permit people to work with different applications using different windows.

music program A computer application that enables users to compose and mix sounds digitally.

N

new media Emerging information technology that combines computers, video/television, and telephony.

O

object-oriented Related to assemblies of information people can readily identify. People can build object-oriented models in computers. Objects can contain many attributes. Some attributes can be inherited from other objects, enabling people to build models more quickly using modules or primitives.

object-oriented programming An approach to computer programming that enables people to build from modules that have attributes that transfer from one program to another.

operating system A computer program that enables people to access files. Permits users to perform functions such as display directories, and search for, copy, or delete files.

P

paper space A representation showing what you will see when you print a document on paper.

parameter A key variable that governs the shape or performance of a model.

personal analogy A similarity or likeness that involves your identification with elements of a problem.

PERT (progress evaluation review technique) A procedure for considering what is completed on a project and relating this to a schedule.

physical interface In computer applications, that which relates to the way people work with the hardware devices.

plug-in An applet, such as Acrobat Viewer, that can work with internet browsers such as MS Internet Explorer or Netscape.

po A positive maybe.

polymorphic tweening A change in morphology (or form) of an object, showing all the different shapes in between.

portal Gateway to nodes or sites on the World Wide Web.

presentation reality The information environment others will perceive. This may involve the layout of pages or drawings you produce on paper, or it might be the sequence of a video production.

principle A fundamental truth upon which others are based.

project management program A computer application that enables users to keep track of resources, activities, and time for doing projects.

R

raster image Computer graphic composed of a bit map indicating which pixels to activate on a computer screen. Sometimes called a bit map image. Enables people to manipulate contrast, color, and texture.

regenerate Natural process to form again or renew. To be spiritually reborn. For example, while computers regenerate images on a screen, human minds recenter and refocus — regenerating mental images in rhythm with life's energy.

renaissance Rebirth; revival. The revival of art, literature, and learning in Europe during the fourteenth, fifteenth, and sixteenth centuries, marking the transition from the medieval to the modern world. The term can also refer to a revival of creativity and understanding stimulated by multimedia computing, empowering individuals to make a transition into a new information age.

rendering program A computer application that enables users to add colors and textures to graphic images.

rotoscope Tracing and cutting out an image to composite for special effects.

S

screen space What you see on a typical computer monitor. The display may vary unless laid out in a standard screen format such as Adobe Acrobat.

script A written document describing a performance. For example, when developed for a multimedia presentation a script can help people integrate different channels of media such as video and audio.

self-actualize To release your inner needs and potential. Psychologist Abraham Maslow places self-actualization at the top of a hierarchy of human needs.

simulation A setup that represents real environments. Users can use digital tools to rehearse what it would be like to do something, like fly an airplane, in reality.

soft prototypes Three-dimensional models built in media space using computer software.

software The computer code that transfers instructions of operating systems and application programs. The program that enables the computer to carry out commands.

spiritual interface That which has to do with a sense of attachment, empowerment, and meaning.

spreadsheet program A computer application that enables users to organize information in rows and columns and do calculations.

stop point The point at which one task ends and another begins. Identifying stop points can make it easier to delegate tasks and transfer information.

storyboard Graphics and text that portray scenes. For example, when used for a multimedia presentation, a storyboard helps people portray the content, composition, and sequence of a presentation.

subdirectory A folder contained within a directory. A computer disk can be divided into many subdirectories.

supply chain Addresses how a project/product moves from the contractor/producer to the end user.

symbolic analogy A similarity or likeness that involves abstract qualities that you can relate from one situation to another.

synchronous At the same time. Online communication engaging participants in interaction at the same time, as in a telephone conversation.

syndrome (From Greek, *syn*, together, and *dromos*, running) Conditions running together. A number of symptoms which together characterize a problem.

T

technographer A technical recorder for computer conferencing. This person uses electronic media to record a group's thinking, so the group can work interactively to collaboratively develop a document.

telecommunications Electronic transfers of information. Can take the form of FAX, e-mail, or, especially, binary transfers of digital files such as programs, formatted documents, drawings, and multimedia.

template A standardized format that presents generic information. For example, this could be as simple as an outline that computer users can add information to and modify for their own purposes.

tool An object that helps the user to work. This may be a traditional tool — such as a pencil — or a digital tool — such as hardware and software for word processing, computer-aided design, animation, and many other applications. Used to optimize efficiency where cost-effective.

toy An object that enables the user to play. Used for pleasure and self-development. The entertainment industry produces a plethora of digital toys.

transfer To convey or send. For example, using electronic media you can transfer text, graphics, images, video, or sound.

transform To change the form or condition of something. For example, digital tools can transform information from paper to electronic media, or from one file format to another.

transformation A change, or visualization of change.

trilogy A discourse consisting of three parts. A set of three.

U

uniform resource locator (URL) The address of a site on the the World Wide Web.

upload To transfer digital information from a local computer to a remote server (often a mainframe computer).

user shell A computer program that provides a graphic user interface making an operating system easier to use. Most user shells incorporate metaphors such as a desktop or windows to help users relate to the interface more intuitively.

V

vector image Computer graphic developed using algorithms that generate the geometry you see on the computer screen. Enables people to manipulate geometry — rotating, mirroring, and changing their scale.

vertical thinking An approach to creative thinking typically involving a linear, logical progression of steps.

video space A representation showing what you will see when you integrate media to produce a video.

virtual office A location-independent organization that links team members through the Internet, Intranets, and Extranets.

virtual private networks (VPNs) Private pathways within a busy public network—making use of the Internet—VPN's are less expensive and offer greater flexibility.

virtual reality A simulation using information to provide what are, in effect, realistic experiences. People can create this simulation by using computer-generated images in media space.

virtual reality markup language (VRML) A file format for three-dimensional web sites.

visual thinking skills Approaches that help you comprehend what you perceive, and help you give expression to patterns. Visual thinking skills enable people to work with representations — drawings, diagrams, models, animation, video, and sound — so they can use multimedia.

visualization The formation a mental image that can help you coordinate your mental and physical activities.

visualization technique Ways to perceptive and comprehend based largely on pattern seeking and pattern recognition.

voice mail Verbal messages transmitted to a computer account using a telephone. These messages can be picked up anytime, from anywhere, using a telephone.

W

wide area network (WAN) Links computers and peripherals together to share information throughout an organization such as a university.

witness To personally experience. Can involve centering on where you are, to become fully aware of your senses.

work flow diagram A diagram linking activities and events and relating them to a time line. Used widely for operations research and project management.

word processing A computer application that enables users to write and edit digital files.

World Wide Web The information environment accessible through Web browsers such as Netscape or MS Explorer.

Index